Cornelis B. Vreugdenhil

Computational Hydraulics

An Introduction

With 122 Figures

Springer-Verlag
Berlin Heidelberg New York
London Paris Tokyo Hong Kong 1989

Professor Dr. Cornelis B. Vreugdenhil
ENR Computer Services for Technology
P.O. Box 1
1755 ZG Petten
The Netherlands

ISBN-13: 978-3-540-50606-5   e-ISBN-13: 978-3-642-95578-5
DOI: 10.1007/978-3-642-95578-5

Library of Congress Cataloging in Publication Data
Vreugdenhil, Cornelis Boudewijn, 1941–
Computational hydraulics: an introduction/
Cornelis B. Vreugdenhil.
Bibliography: p.   Includes index.
ISBN-13: 978-3-540-50606-5
1. Hydraulics—Mathematical models.   2. Hydraulics—Mathematics.
I. Title.
TC163.V85   1989

Typesetting: Macmillan, India;

2161/3020-543210 – Printed on acid-free paper

Contents

Chapter 1
Introduction

What is Computational Hydraulics?

Computational hydraulics is one of the many fields of science in which the application of computers gives rise to a new way of working, which is intermediate between purely theoretical and experimental. It is concerned with simulation of the flow of water, together with its consequences, using numerical methods on computers. There is not a great deal of difference with computational hydrodynamics or computational fluid dynamics, but these terms are too much restricted to the fluid as such. It seems to be typical of practical problems in hydraulics that they are rarely directed to the flow by itself, but rather to some consequence of it, such as forces on obstacles, transport of heat, sedimentation of a channel or decay of a pollutant. All these subjects require very similar numerical methods and this is why they are treated together in this book. Therefore, I have preferred to use the term computational hydraulics. Accordingly, I have attempted to show the wide field of application by giving examples of a great variety of such practical problems.

Purpose of the Book

It is getting a normal situation that an engineer is required to solve some engineering problem involving fluid flow, using standard and general-purpose computer programs available in many organizations. In many instances, the software has been designed with the claim that no numerical or computer-science expertise is needed in using them. However, it is not uncommon that wrong results are obtained due to lack of insight and, unfortunately, such results are not always easily recognized. Therefore, the user should have a basic knowledge of what is going on inside the programs and in particular should be able to evaluate the numerical results with respect to physical relevance and numerical reliability. The first purpose of this book is to give insight into the type of methods used in such computer programs and the ways of critically evaluating the results. In actual practice, there may be a need to consult specialists. With the background of this book, the engineer should then be able to ask the relevant questions. Conversely, the specialist should have some knowledge of the field of applications. In other words, there should be some common ground between him and the users. The second purpose of this book is to provide him with the necessary information on the application side.

The book is based on courses for third-year hydraulic-engineering students at the Delft University of Technology (The Netherlands). It has also, in part or in full, been used for various other courses.

Approach

The general approach in this book is to discuss a group of applications starting from the mathematical formulation, through numerical methods of solution to examples of practical interest. The emphasis is not on a great variety of different methods, but rather on the common aspects between various cases. Therefore, only a very small number of numerical methods are discussed which are considered to be typical and well-proven in practice.

The main theme is the analysis of the numerical behaviour, discussed in such a way that you should be able to apply it more generally to any similar numerical method you might meet. In many cases the behaviour of the methods can be analysed by solving the linearized equations for special but typical cases. The numerical effects can be quantified in terms having a physical interpretation such as propagation speed or damping rate. It is then assumed that these estimates can also be used to indicate the numerical effects in the general nonlinear case. Examples show that this is often true.

As a consequence of this approach, the linearized and simplified equations get a great deal of attention. Moreover, to keep things simple, I often introduce a particular numerical method for the linear equations and assume that you can generalize it yourself. You should not draw the conclusion that these simplified equations are the ones used in actual applications. On the contrary, the complete equations with all the complications required for the problem at hand should be used. Actually, in many commercially available computer programs, the most complete form of the equations is built in and the user may switch off those terms that are not needed in a particular case.

Subjects

In the choice of subjects, I have taken into account the most common types of applications. As far as physics are concerned, hydrodynamics is discussed for one-dimensional and two-dimensional shallow-water flow (rivers and seas). Boundary-layer flow is briefly discussed with the main purpose of establishing a link with more advanced two- and three- dimensional hydrodynamics, which are outside the scope of this book. Two-dimensional potential flow is treated, with groundwater as its main application. Transport of passive substances (salt), decaying pollutants and suspended sediment are examples of convection-diffusion problems. Diffusion is also discussed in other cases such as groundwater flow, coastline development and consolidation of soil.

In a mathematical sense, most problems are described in terms of ordinary or partial differential equations. I have included flow and transport processes involving not more than two independent variables. The only exception to this is two-dimensional flow in shallow water, which is such an important field of application that it could not be disregarded. Other higher-dimensional methods have not been included as they are still too much in a research phase. On the other hand, in such cases one often takes recourse to lower-dimensional analyses (such as discussed here) to have at least a partial understanding of what is happening.

A discussion of two- and three-dimensional flow of a viscous or turbulent fluid (based on the Navier Stokes equations) has been omitted for various reasons. The required level would be higher than what is generally assumed in this book. Perhaps more importantly, there are several good books available on this subject (e.g. Roache 1976, Chung 1978, Peyret & Taylor 1983), which unfortunately is not the case for the majority of the material in the present book.

From a numerical point of view, almost exclusively finite-difference methods are discussed. Finite elements are treated briefly for elliptic problems (potential flow), with the main purpose to show the similarity with finite-difference methods. A more extensive discussion of finite elements is not needed as there is a vast literature available; see e.g. Baker (1983), Chung (1978), Pinder & Gray (1977), Taylor & Hughes (1981).

An important theme in the book is the similarity between various seemingly unrelated phenomena. It is quite useful to be aware of such a similarity, because you may borrow ideas from some well-developed field of application and translate them to another, less common, field. A good example is the similarity between the shallow-water (Saint Venant) equations and those for compressible aerodynamics. For this reason, I have included examples which perhaps do not belong to the field of hydraulics strictly speaking, such as coastline dynamics and consolidation of soil.

Most of the examples in this book have been specially developed for the purpose of demonstration. As a consequence, they tend to be somewhat schematic. The reason is that many real-world examples have some confusing details which make it difficult to bring out a specific point. The consequence of this approach is unavoidably that your own practical applications tend to look much more complicated than the examples. However, I am convinced that once you have seen the basic phenomena in simple cases, you will be able to recognize them more easily in your practical problems which, in the majority of cases, may be reduced to one of the simple examples. This will help you in finding an explanation or a remedy. For the same reason, I have abandoned the idea of including case studies. You can find these in abundance in the various periodicals on the subject.

Some of the examples have been prepared for this book by some of my colleagues. The contributions by Nico Booij and Marius Sokolewicz from the Delft University of Technology and Teun Burgers and Daan Bakker from ENR Computer Services for Technology in this respect have been most helpful. Other results have been previously published, sometimes together with my colleagues. These are accounted for in the references.

What Not to Expect

You should not expect to be a fully trained developer of hydraulic computer programs after having studied this book. There is much more to well designed, well tested and well documented software than discussed here. This does not exclude that you may write some simple computer programs for your own use on your personal computer (actually, that may be a very good exercise), but do not claim those to be professional software.

Previous Knowledge

The knowledge of mathematics that is presupposed is modest and does not exceed basic calculus including linear algebra and complex numbers. It is quite helpful if you have some prior knowledge of ordinary and (in particular) partial differential equations; however, this is not assumed and the basic ideas are explained in the text. As far as numerical methods are concerned, it is helpful if you know how linear systems of equations and ordinary differential equations are solved.

On the hydraulics side, a general familiarity with the equations of flow and transport is assumed. However, to keep the book self-contained, I include (short) derivations of the equations.

Chapter 2
Water Quality in a Lake

In the next few chapters, the main ideas of this book are introduced in a very simple example, where most of the equations can be solved analytically. The type of model is called a "box" model and it is governed by ordinary differential equations. In the simplest case, there is just one first-order equation and the system is accordingly called a first-order system.

2.1 Mathematical Formulation

Consider a small lake in which some sort of waste is discharged. Part of the waste material is degraded by biological activity; moreover, the polluted water is flushed by a river, entering and leaving the lake. The level of pollution is expressed in terms of mg BOD/l (milligram biochemical oxygen demand per litre), which can be treated as a concentration. A conservation equation for it can be set up. Suppose that the water in the lake is well mixed as a consequence of, e.g., wind action. Then the concentration in the lake can be represented reasonably well by one single number, indicated by c.

The first step in setting up a conservation equation is choosing a control volume. The choice is obvious in this case: it is the entire lake.

A conservation equation always has the form

$$\text{input} - \text{output} = \text{storage} \tag{2.1}$$

For the case of BOD, we get:

– input:
1. input from the river $Q_i c_i$ where c_i is the concentration of BOD of the river water and Q_i the incoming discharge;
2. the amount L discharged into the lake (unit: mg BOD/s)

– output:
1. the material leaving the lake into the river; as the lake is well mixed, it is

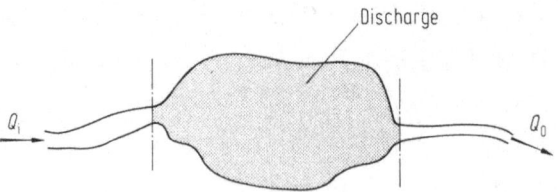

Discharge

Fig. 2.1 Control volume.

reasonable to assume that the concentration at the outlet is the same as the average concentration in the lake, so the outflow is $Q_o c$;

2. the concentration also decreases by biological degradation; this is a negative term in the conservation equation. The "outflow" per unit volume is c/T_r, where T_r is a time scale for the degradation. If the volume is V, the total degradation is Vc/T_r.

– storage:
the rate of change of the amount of waste material in the lake. The instantaneous amount is Vc, so the rate of change is $d(Vc)/dt$.

Collecting all terms, we get the equation:

$$\frac{d}{dt}(Vc) = Q_i c_i + L - Q_o c - Vc/T_r \tag{2.2}$$

If the inflow and outflow are equal and (consequently) the volume is constant, this gives

$$V\frac{dc}{dt} = Q_i(c_i - c) + L - Vc/T_r \tag{2.3}$$

In order to solve this equation, some additional information is needed:

– the initial concentration;
– the magnitude of the river discharge;
– the concentration of the waste material in the inflow;
– the magnitude of the discharge L.

The situation is in equilibrium if $dc/dt = 0$ and from eq. (2.3) it can be seen that this is the case if

$$c = c_e = (L + Q_i c_i)/(Q_o + V/T_r) \tag{2.4}$$

This is the equilibrium concentration. Equation (2.3) can now be written much more conveniently as

$$\frac{dc}{dt} + \frac{c}{T} = \frac{c_e}{T} \tag{2.5}$$

in which

$$T = V/(Q_i + V/T_r) \tag{2.6}$$

The role of the time scale T is explained below; its extremes are:

– no river flow; $T = T_r$ which is the biological degradation time scale;
– no biological degradation $(T_r \to \infty)$: $T = V/Q_i$ which is the flushing time.

If the initial condition is

$$c = c_0 \tag{2.7}$$

the analytical solution can be easily found (please check);

$$c(t) = c_e + (c_0 - c_e)e^{-t/T} \tag{2.8}$$

This is shown in Fig. 2.2. The equilibrium situation is approached asymptotically and the time scale T indicates how fast. More precisely, the *relaxation time T* is the time in which the deviation from the equilibrium concentration is reduced to e^{-1} times its original value. At a few times T, the influence of the initial condition has almost vanished (to 0.14 at $t = 2T$ and 0.05 at $t = 3T$).

As an example, take some numerical values:

$$Q = 1\,\mathrm{m^3/s}$$

$$V = 6\ 10^5\,\mathrm{m^3}$$

$$T_r = 3\,\text{days} = 3.3\ 10^5\,\mathrm{s}$$

$$c_i = 5\,\mathrm{mg/l}$$

Then you will find

$$c_e = 37.5\,\mathrm{mg/l}$$

$$T = 2.5\,\text{days}$$

This time scale is an internal one to the process. There are other time scales describing the variation of the external conditions. Both the "equilibrium" concentration and the relaxation time may vary at an "external" time scale. This means that at every instant the system responds to the difference between the actual concentration and the prevailing equilibrium value, with the prevailing time scale.

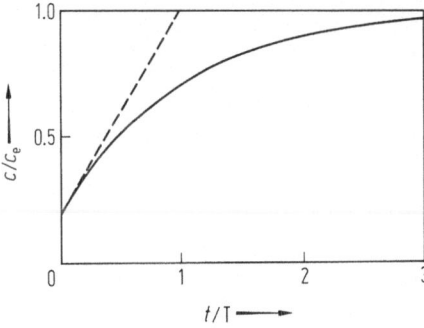

Fig. 2.2 Solution of first-order equation.

If the time scale of external variation T_e is large compared with the relaxation time T, the system is able to adjust itself quickly and it will behave "quasi-steadily". In the case of fast external variations, the system is lagging too far behind to follow the peaks. Both cases are shown in Fig. 2.3.

For time dependent systems, it is often difficult to get good values for the initial conditions. Even if you make a good guess, it may be wrong. Fortunately, in many cases, including the present first-order system, the influence of a wrong initial condition gradually fades away. This is illustrated in Fig. 2.4, where the same situation is approached from two different initial conditions. You will not be surprised that the relaxation time again indicates how quickly this convergence takes place.

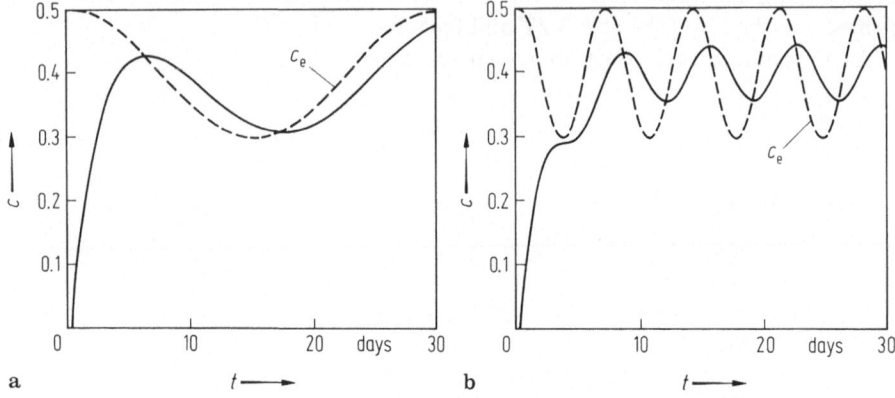

Fig. 2.3 Response of first-order system to external variations.

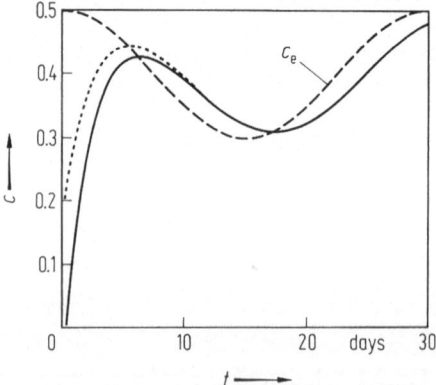

Fig. 2.4 Two different initial conditions converge to the same solution.

2.2 Exercises

These exercises show that many other problems in hydraulics and hydrology (and, as a matter of fact, in other fields of physics) are also formulated as first-order systems. In each case, show this by deriving the differential equation. Give expressions for the equilibrium values and the relaxation time scales.

1. To determine the run-off of rain water from inaccessible areas into a river, one often uses a very simple conceptual model, supposed to represent the physical system in an overall sense. It consists of a storage basin with surface area F and water level $h(t)$, in which the rainfall with intensity $r(t)$ per unit area is collected. The run-off Q into the river is assumed to be proportional to the difference $h - h_0$ with a constant of proportionality q. The unknown parameters F, q, h_0 are estimated using observed records of rainfall and run-off. The variable in the model is $Q(t)$.
2. A sand grain with mass m falls within stagnant water. Suppose it has an added mass m_a due to acceleration of the displaced water. At low values of the fall velocity $w(t)$, the resistance force is proportional to the fall velocity and the cross-sectional area A of the sand grain. The variable in the model is $w(t)$.

Chapter 3
Numerical Solution for Box Model

3.1 Principle

For simple cases the differential equation for the box model describing the water quality problem of Chapter 2 can be solved analytically. If the inflow or the waste discharge varies in an arbitrary way, this is no longer so. Moreover, in many applications of box models the equations will not be so nicely linear. In general, you will need numerical techniques and these can be illustrated very well for the water quality example.

Suppose that c_e and T are known, possibly as functions of time but for simplicity let us take them as constants. If the concentration c_j at time t_j is known, eq. (2.4) gives the rate of change of c. Assuming this rate to remain constant during a small time interval Δt, an estimate for the concentration after that interval is:

$$c_{j+1} \approx c_j + \Delta t \left. \frac{dc}{dt} \right|_{t=t_j} = c_j + \frac{\Delta t}{T}(c_e - c_j) \qquad (3.1)$$

Starting at the initial concentration c_0 at time $t = t_0$ the behaviour of the concentration can be simulated with time intervals Δt. It is intuitively clear that this is the more accurate if the time intervals or time steps are small. Figure 3.1 illustrates this. One of the main questions addressed in this book is how small should the time steps be.

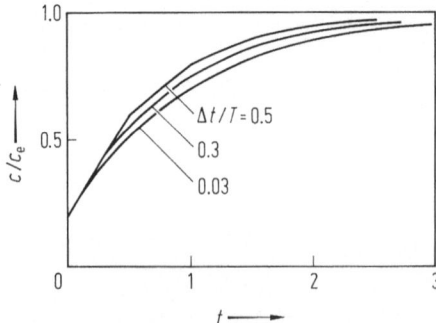

Fig. 3.1 Numerical solution of the water-quality problem for different values of the time step Δt.

3.2 Stability and Accuracy

In this simple example, the influence of the time step Δt can be analysed by solving the finite-difference equation (3.1) exactly. Note that this is not possible in many more realistic applications, but as a guideline, this analysis is useful. You can easily check that the solution of eq. (3.1) is:

$$c_j = c_e + \left(1 - \frac{\Delta t}{T}\right)^j (c_0 - c_e) \tag{3.2}$$

If you compare this with the analytical solution (2.7) it is apparent that both solutions are different. This is a normal situation and you should not be concerned about it. Looking more closely, both solutions can be seen to be near one another under certain conditions, and it is exactly these conditions you would like to know. Apparently the dimensionless parameter $\Delta t/T$ determines the behaviour of the method. There are four cases:

(i) if $\Delta t/T \ll 1$ then $\qquad \left(1 - \dfrac{\Delta t}{T}\right)^j \approx e^{-j\Delta t/T}$

and the two solutions agree reasonably well.

(ii) if $\Delta t/T < 1$ but not $\ll 1$, then $(1 - \Delta t/T)^j$ is a poor approximation of $\exp(-t/T)$, so the numerical solution is inaccurate.

(iii) if $1 < \Delta t/T < 2$ then $1 - \Delta t/T < 0$ so the numerical solution oscillates and is certainly inaccurate. This case can be seen in Fig. 3.2.

(iv) if $\Delta t/T > 2$ then $1 - \Delta t/T < -1$ so the numerical solution not only oscillates but it even grows exponentially, which clearly deviates from the analytical

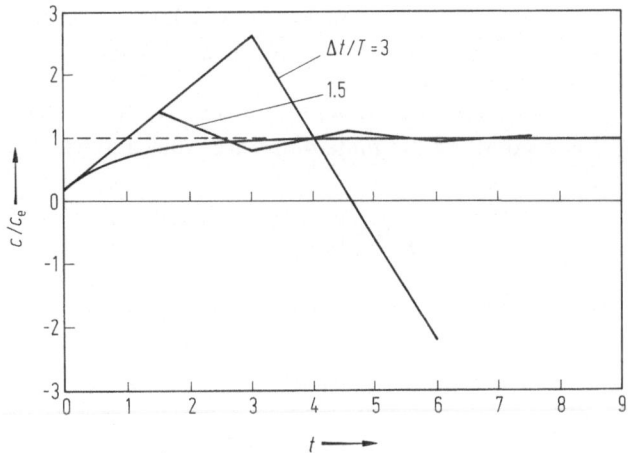

Fig. 3.2 Oscillating and unstable solutions.

solution (again see Fig. 3.2). This is called instability and it is, unfortunately, a rather common problem in numerical methods.

This analysis shows that a condition for stability is

$$\Delta t/T < 2 \tag{3.3}$$

but this does not imply accuracy. For the latter, $\Delta t/T$ must be much smaller; just how small can be found by comparing the numerical and analytical solutions more quantitatively. For example, a "numerical relaxation time" can be defined as the time $j \Delta t$ at which

$$(1 - \Delta t/T)^j = e^{-1}$$

which gives

$$j \frac{\Delta t}{T} = \frac{-\Delta t/T}{\ln(1 - \Delta t/T)} \approx (1 + \tfrac{1}{2}\Delta t/T)^{-1} \tag{3.4}$$

For an accurate solution, this should be close to unity. Equation (3.4) gives the opportunity to choose the time step such that you meet a certain predetermined accuracy criterion.

There are more factors influencing the time step. It should be small not only relative to the internal time scale, but also to the external one T_e, say 5 or 10% of it. In case of rapid variations, this is a heavier requirement than the one resulting from eq. (3.4). For slow variations, eq. (3.4) may be more restricting, but in that case it is not so important that the relaxation time is reproduced very accurately, so a relatively large error may be allowed. In any case, the stability requirement should be observed.

The preceding analysis considers the error in the numerical solution and in the remainder of this book you will find similar estimates for more complicated equations. Unfortunately, it is not always possible to perform such an analysis. There is an alternative that is always possible, even for nonlinear equations. It gives (an estimate of) the error in the equation, not in the solution. It is, therefore, not the information you really need, but at any rate it is of some use to know which equation you have actually been solving.

In this analysis, the (unknown) solution is developed into a Taylor series with respect to the "base point" of the finite-difference equation, which is the time t_j. Then

$$(c_{j+1} - c_j)/\Delta t = \left.\frac{dc}{dt}\right|_j + \tfrac{1}{2}\Delta t \left.\frac{d^2c}{dt^2}\right|_j + \cdots \tag{3.5}$$

Substitute this into eq. (3.2):

$$\left.\frac{dc}{dt}\right|_j + c_j/T = c_e/T - \tfrac{1}{2}\Delta t \left.\frac{d^2c}{dt^2}\right|_j + \cdots \tag{3.6}$$

This agrees with the original differential equation except for the last term, which is called the *truncation error*. In this case it is of the first order in Δt, so if you reduce Δt by a factor of $\tfrac{1}{2}$, the error is reduced by the same factor. By using eq. (3.1), the magnitude of the truncation error can be determined:

$$-\tfrac{1}{2}\Delta t \,\frac{d^2 c}{dt^2} \approx \tfrac{1}{2}\Delta t(c_e - c)/T^2 \tag{3.7}$$

In comparison with the term $(c - c_e)/T$ in the equation, this gives a relative error of $\tfrac{1}{2}\Delta t/T$ which happens to agree with the estimate in eq. (3.4). Note that the truncation error vanishes if the solution becomes steady. Then the value of the time step is no longer important. This also agrees with the previous analysis, because in a steady state it is no longer necessary to reproduce the relaxation time.

3.3 Example

In the lake of section 2.1 a great quantity of BOD is discharged during one day. The corresponding equilibrium concentration would be 100 mg/l. Suppose that a maximal concentration of 50 mg/l is allowed, will it be exceeded? As the relaxation time is 2.5 days, you suspect that it will not; however, to be sure, a numerical simulation can be made. The time step should be chosen such that the result is reliable.

The external time scale is 1 day (the response to the sudden changes is governed by the relaxation time). After one day—the most critical time—the concentration should be computed with an accuracy of, say, 1%, i.e. $(1 - \Delta t/T)^j$ should not deviate more than 1% from $\exp(-t/T)$ for $t = j\Delta t = 1$ day. If you experiment a little with $\Delta t/T$, a value of 0.05 is found to be satisfactory, so $\Delta t = 3$ h. This is also reasonably small with respect to T_e.

To make the simulation, an initial condition is needed. Suppose $c_0 = 30$ mg/l. The computed concentration is shown in Fig. 3.3. It is seen that the critical value of

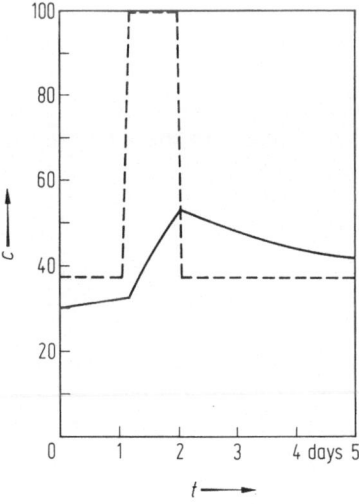

Fig. 3.3 Concentration history after a sudden release.

50 mg/l is hardly exceeded; this is, however, somewhat dependent on the initial value.

3.4 Implicit Method

The method of section 3.1, known as Euler's method, is restricted by the stability criterion (3.3) even if the solution varies quite slowly. This is typical of explicit methods. It can be avoided by approximating the differential equation at time $t_j + \theta \Delta t$ with $0 \leqslant \theta \leqslant 1$. At that time,

$$(c_e - c)/T \approx \theta(c_{e\,j+1} - c_{j+1})/T_{j+1} + (1 - \theta)(c_{e\,j} - c_j)/T_j$$

Then the finite-difference equation becomes

$$\frac{c_{j+1} - c_j}{\Delta t} = \theta(c_{e\,j+1} - c_{j+1})/T_{j+1} + (1 - \theta)(c_{e\,j} - c_j)/T_j \tag{3.8}$$

As the unknown value c_{j+1} now occurs at both sides of the equation, this is called an *implicit* method: an algebraic equation has to be solved at each time step. In this case, it happens to be a single linear equation, but this may be much more complicated in other cases.

In an exercise (section 3.5), it is derived that the truncation error in this case is

$$(\theta - \tfrac{1}{2})\Delta t\,\frac{d^2 c}{dt^2} + O(\Delta t^2) \tag{3.9}$$

so it is still of the first order, but generally smaller than for Euler's method (which, incidentally, is included as a special case for $\theta = 0$).

If the coefficients are constant, the solution of eq. (3.8) can again be found:

$$c_j = c_e + \left\{\frac{1 - (1 - \theta)\Delta t/T}{1 + \theta \Delta t/T}\right\}^j (c_0 - c_e) \tag{3.10}$$

The factor between curly brackets should be smaller than unity in absolute value for stability; it is called the *amplification factor*. It is not difficult to check that this is so for any value of Δt if

$$\tfrac{1}{2} \leqslant \theta \leqslant 1 \tag{3.11}$$

The importance of this is that the time step is no longer restricted by stability. It can be chosen purely on accuracy grounds. From eq. (3.10), a numerical relaxation time can again be determined as in eq. (3.4).

There exist many more numerical methods to solve ordinary differential equations, both explicit and implicit. See any good book on numerical methods, e.g. Gear, 1971. They are not discussed here, because most of the more sophisticated methods are not used in solving partial differential equations, as in the remainder of this book.

3.5 Exercises

1. Show that the truncation error of the implicit scheme is given by eq. (3.9). Hint: take the base point for the Taylor series at time $t + \theta \Delta t$.
2. What time step could be used for the example of section 3.3 if the implicit scheme is applied with $\theta = \frac{1}{2}$? And if $\theta = 1$?

Chapter 4
Transport of a Dissolved Substance

4.1 Mathematical Formulation

Suppose that a certain quantity of some substance is discharged into a river, how is it going to be transported? For the time being, diffusion is neglected. This is not very realistic, but you will find more about that in later chapters. The consequence is that the substance is carried with the packet of water in which it was discharged, see Fig. 4.1. If the discharge takes a time δt, the length of the packet is $u\delta t$, where u is the flow velocity. After a time t, the entire packet has been transported over a distance ut. In the meantime, the substance may have decayed with a relaxation time T, just as in Chapter 2 (now only due to degradation, as there is of course no in- or outflow for the packet of water). In fact, you could use the same formulation as in that chapter in a frame of reference moving with the flow velocity. However, that approach cannot be easily generalized to more complicated cases. Therefore, it is better to consider the mass balance for a (stationary) elementary control volume shaped as a slice which covers the entire river cross-section A over a length Δx. During a time interval Δt (Fig. 4.2),

The input is $\qquad Q(x, t)c(x, t)\Delta t$

output $\qquad Q(x + \Delta x, t)c(x + \Delta x, t)\Delta t$

decay $\qquad A\Delta x c(x, t)\Delta t/T$

storage $\qquad A(x, t + \Delta t)c(x, t + \Delta t)\Delta x - A(x, t)c(x, t)\Delta x$

Actually, some of the quantities should be taken at an intermediate time or location, e.g. $c(x + \frac{1}{2}\Delta x, t)$ in the input; however, this does not change anything in taking the limit for Δx and $\Delta t \to 0$ (please check).

$$\frac{\partial}{\partial t}(Ac) + \frac{\partial}{\partial x}(Qc) + Ac/T = 0 \qquad (4.1)$$

In a similar way the balance of water for the same control volume can be considered; this results in

$$\frac{\partial A}{\partial t} + \frac{\partial Q}{\partial x} = 0 \qquad (4.2)$$

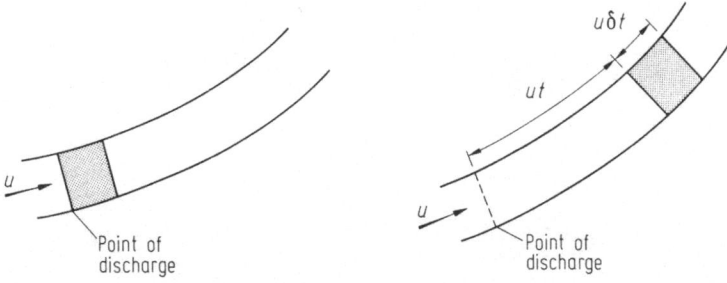

Fig. 4.1 Displacement of a packet of water with dissolved substance; left directly after discharge, right after a time t.

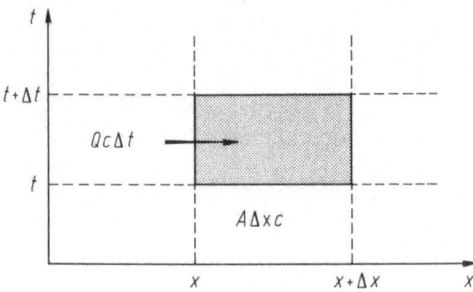

Fig. 4.2 Control volume for mass balance.

where Q is the discharge of water. Performing the differentiation in eq. (4.1) and using eq. (4.2) you will find

$$\frac{\partial c}{\partial t} + u\frac{\partial c}{\partial x} + \frac{c}{T} = 0 \tag{4.3}$$

This type of equation occurs in many other types of physical problems; it is called the *simple-wave* equation for reasons explained below.

The behaviour of the solution can be demonstrated most clearly if there is no decay ($T \to \infty$). Then eq. (4.3) shows, as you would expect, that the concentration in a packet of water is constant. If an observer moves with the flow velocity (i.e. along a track with $dx/dt = u$) he will observe a constant concentration (Fig. 4.3a). The magnitude of this concentration is the same as that at the starting point of the track. If this is at time $t = 0$ downstream of the point of discharge, you have to know the concentration there, so you need an initial condition at each point of the region at time $t = 0$. If the track starts at the discharge point at time $t > 0$, you need to know the concentration, which is a boundary condition. In this case, the concentration at the point of discharge can be determined from a mass balance again. If no water is discharged and if the original concentration in the river water is zero, the mass flow of the substance Qc just downstream should equal the

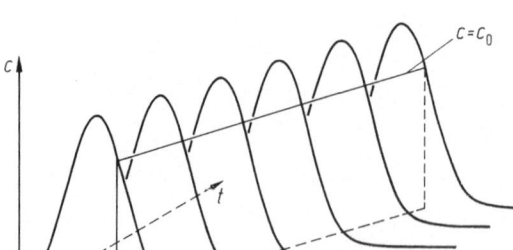

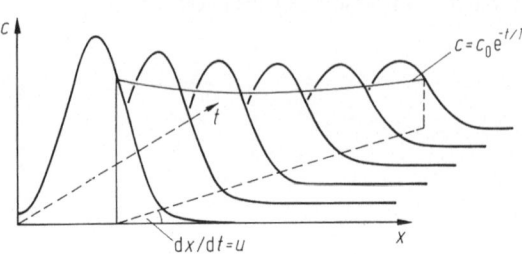

Fig. 4.3 Behaviour of solution without and with decay.

amount discharged $F(t)$, so

$$c(0, t) = F(t)/Q(0, t) \tag{4.4}$$

Here it has been assumed that there is complete mixing at the discharge point, which may not be realistic for large rivers.

If there is decay (T finite, see Fig. 4.3b) the situation is similar. The observer moving with the flow velocity does not see a constant concentration, but one decaying exponentially with relaxation time T. The consequences for initial and boundary conditions are the same.

The tracks, along which the concentration remains constant in the first case or decays exponentially in the second, are called *characteristics*. In this case, a characteristic passes through each point (x, t) and it coincides with the track of a water packet:

$$dx/dt = u \tag{4.5}$$

Later, other types of characteristics will be met, which do not have anything to do with the flow velocity. The characteristics describe some characteristic property of the solution (hence the name). The type of solution in this case is just a "simple" wave moving with the characteristic velocity u (Fig. 4.4) and being damped either or not.

If the number of characteristics at each point is equal to the order of the partial differential equation (as it is here; you may think that this is obvious, but it is not), the system is called *hyperbolic*. Later parabolic and elliptic cases will be met. Usually, hyperbolic equations have something to do with wave phenomena.

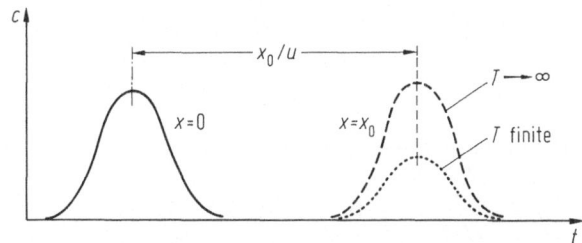

Fig. 4.4 Concentration as a function of time at locations $x = 0$ and $x = x_0$ with or without decay

4.2 Numerical Solution

The simple-wave equation (4.3) can be solved numerically by discretisation. Now not only time steps Δt are introduced, but also space is discretized into a grid with grid size Δx (Fig. 4.5). The value of the concentration at time $n\Delta t$ and place $j\Delta x$ is indicated by c_j^n. For the time being, suppose that $u = $ constant and there is no decay.

The partial derivatives can now be approximated by finite differences in the point $(j\Delta x, n\Delta t)$ as:

$$\left.\frac{\partial c}{\partial t}\right|_j^n \approx \frac{c_j^{n+1} - c_j^n}{\Delta t} \tag{4.6}$$

$$\left.\frac{\partial c}{\partial x}\right|_j^n \approx \frac{c_{j+1}^n - c_j^n}{\Delta x} \tag{4.7}$$

Both are forward differences. For the spatial derivative, it is more useful from an

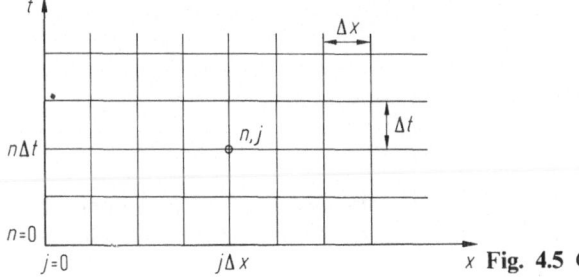

Fig. 4.5 Grid for finite-difference method.

accuracy point of view to use central differences:

$$\frac{\partial c}{\partial x} \approx \frac{c_{j+1}^n - c_{j-1}^n}{2\Delta x} \tag{4.8}$$

Combining eqs. (4.6) and (4.8) gives the difference equation

$$\frac{c_j^{n+1} - c_j^n}{\Delta t} + u \frac{c_{j+1}^n - c_{j-1}^n}{2\Delta x} = 0 \tag{4.9}$$

or

$$c_j^{n+1} = c_j^n - \frac{u\Delta t}{2\Delta x}(c_{j+1}^n - c_{j-1}^n) \tag{4.10}$$

Given the initial values at $t = 0$ $(n = 0)$ for all grid points, it is possible to compute values at time $t = \Delta t$ $(n = 1)$ in all grid points except the boundary points $j = 0$ and $j = J$. At the former, a boundary condition is given. At the boundary $j = J$, which is an artificial one, eq. (4.10) cannot be used as it would require a point outside the region. Therefore, a variant of eq. (4.10) using backward differences is used:

$$\frac{c_J^{n+1} - c_J^n}{\Delta t} + u \frac{c_J^n - c_{J-1}^n}{\Delta x} = 0 \tag{4.11}$$

In this way the entire time level $n = 1$ can be computed. Then the process can be repeated for the next time level, as far as needed. Each new value of the concentration is computed explicitly from values at the preceding time level; that is why this type of finite-difference schemes is called *explicit*.

   An example is given in Fig. 4.6 together with the analytical solution, which can be determined easily using the theory of characteristics from the previous section.

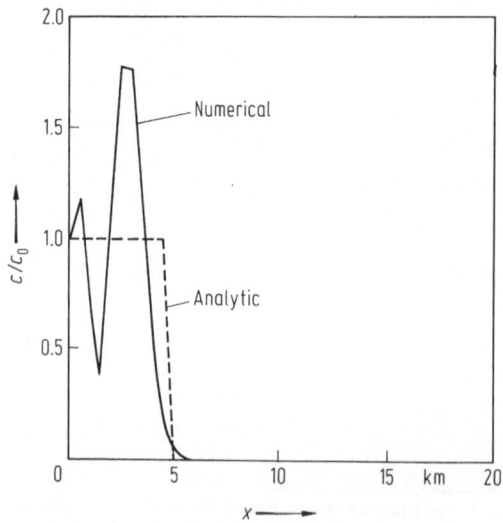

Fig. 4.6 Salt transport from a continuous release at $x = 0$; situation at time 3 hours.

The situation is a continuous discharge of a conservative (non-decaying) substance, like salt, at a point $x = 0$ starting at time $t = 0$. The concentration directly downstream of the point of discharge is called c_0. The flow velocity is $u = 0.5$ m/s. After 3 hours, the salt would have extended over a length of about 5 km. As no diffusion is taken into account, the front remains sharp. The numerical parameters are: $\Delta x = 500$ m and $\Delta t = 500$ s. Apparently something is wrong: you find an oscillating solution that gets worse and worse. This can be explained as follows. Suppose you would have had an initial condition of the form

$$c(x, 0) = c_0 \cos(kx) \tag{4.12}$$

For ease of analysis, we will write this as

$$c(x, 0) = c_0 \exp(ikx) \tag{4.13}$$

The preceding case can be obtained by just taking the real part. In the grid points, the initial values are

$$c_j = c_0 \exp(ij\xi) \tag{4.14}$$

in which

$$\xi = k\Delta x = 2\pi \Delta x / L$$

and L is the wave length. If you introduce this into the difference equation (4.10), each term contains a factor $\exp(ij\xi)$, so the new solution has a form similar to (4.13):

$$c_j = c_0\{1 - \tfrac{1}{2}\sigma(e^{i\xi} - e^{-i\xi})\} \exp(ij\xi) = c_0(1 - \sigma i \sin \xi) \exp(ij\xi) \tag{4.15}$$

where $\sigma = u\Delta t / \Delta x$. The numerical solution has been multiplied during this time step by a complex factor called the *amplification factor*

$$\rho = 1 - \sigma i \sin \xi \tag{4.16}$$

which means that the amplitude is multiplied by $|\rho|$ and there is a phase shift by $\arg(\rho)$. During the next time step the same thing happens. Now, the amplification factor is greater than unity in absolute value for any ξ and σ, so the amplitude of the numerical solution will increase (even exponentially) whereas it should remain constant. This is called instability. In the example of Fig. (4.6) we did not have a cosine function as an initial condition, but you can develop any initial function into a Fourier series consisting of cosine functions with various wave lengths. Each of these will grow according to (4.15) and the combined effect is what you see in the figure. The finite difference scheme (4.10) is therefore useless.

The analysis of stability in terms of Fourier series, as shown, is due to Von Neumann. You will see in the next chapter that there are other finite-difference schemes that are stable. Apparently, a general condition for stability is that the amplification factor should be less than unity in absolute value:

$$|\rho| \leqslant 1 \tag{4.17}$$

for any value of the wave length, or rather of the parameter ξ which contains the ratio of grid size to wave length.

4.3 Exercises

1. Consider a river with a cross-sectional area of 200 m². The discharge is constant 100 m³/s. At a point $x = 0$ a quantity of 100 kg/s of salt is discharged from $t = 0$ to $t = 15$ min. The river water originally does not contain any salt. If you neglect any diffusion, what will be the concentration

   − at $x = 2$ km and $t = 1800$ s;
   − at $x = 1$ km and $t = 2400$ s?

2. In the unstable computation shown in Fig. 4.6 you notice a "wave" with wave length about 3 km. Compute the growth rate per time step of this wave. This gives you an impression of how the instability works.

Chapter 5
Explicit Finite-Difference Methods

5.1 Two-Level Methods

If you restrict yourself to using the same grid points as those in section 4.2, a general explicit method can be written as

$$\frac{c_j^{n+1} - \frac{1}{2}\alpha(c_{j+1}^n + c_{j-1}^n) - (1-\alpha)c_j^n}{\Delta t} + u\frac{c_{j+1}^n - c_{j-1}^n}{2\Delta x} = 0 \tag{5.1}$$

where α is a free parameter that can be manipulated for stability and accuracy. An exercise for this chapter shows that this is indeed a general method for approximation of the simple-wave equation without a decay term. If $\alpha = 1$, you get the method of Lax; therefore the general case can be called a modified Lax method. The difference equation used in section 4.2 is a special case $\alpha = 0$.

The boundaries can be treated the same way as in section 4.2. At $t = 0$ an initial condition is given, which is sufficient to start the numerical solution. At the upstream boundary $x = 0$ (if $u > 0$) a boundary condition is supposed to be available. At the downstream border, no boundary condition is available and you can use the "upstream" difference equation (4.11).

To study the stability of this method, the Von Neumann method is used again. Using the same approach as in section 4.2, the amplification factor is found to be

$$\rho = 1 - \alpha + \alpha \cos \xi - \sigma i \sin \xi \tag{5.2}$$

where $\sigma = u\Delta t/\Delta x$ and $\xi = 2\pi\Delta x/L$. The smallest possible grid size is obviously $\Delta x = 0$. On the other hand, the largest possible value is $\Delta x = L/2$, as it is not possible to represent a "wave" with less than two grid points per wave length. Conversely, the smallest possible wave length that can be represented on a certain grid is $2\Delta x$. Anyway, the conclusion is that

$$0 \leqslant \xi \leqslant \pi \tag{5.3}$$

The requirement for stability is that

$$|\rho| \leqslant 1 \tag{5.4}$$

for all ξ satisfying eq. (5.3). Elaboration of eq. (5.4) gives:

$$|\rho|^2 = \{1 + \alpha(\cos \xi - 1)\}^2 + \sigma^2 \sin^2 \xi \leqslant 1$$

$$2\alpha(\cos \xi - 1) + \alpha^2(\cos \xi - 1)^2 + \sigma^2(1 - \cos^2 \xi) \leqslant 0$$

Dividing this by $(\cos \xi - 1) \leqslant 0$:

$$2\alpha + \alpha^2(\cos \xi - 1) - \sigma^2(1 + \cos \xi) \geqslant 0$$

This is a linear function of $\cos \xi$, so if the inequality is satisfied for the two extreme values ± 1, it is for the intermediate values as well. This gives

$$\cos \xi = 1 \rightarrow 2\alpha - 2\sigma^2 \geqslant 0 \rightarrow \sigma^2 \leqslant \alpha$$

$$\cos \xi = -1 \rightarrow 2\alpha - 2\alpha^2 \geqslant 0 \rightarrow 0 \leqslant \alpha \leqslant 1$$

The two conditions combined give the stability condition for the modified Lax scheme:

$$\sigma^2 \leqslant \alpha \leqslant 1 \tag{5.5}$$

The meaning of this is that you can choose α and Δx and then Δt is limited by

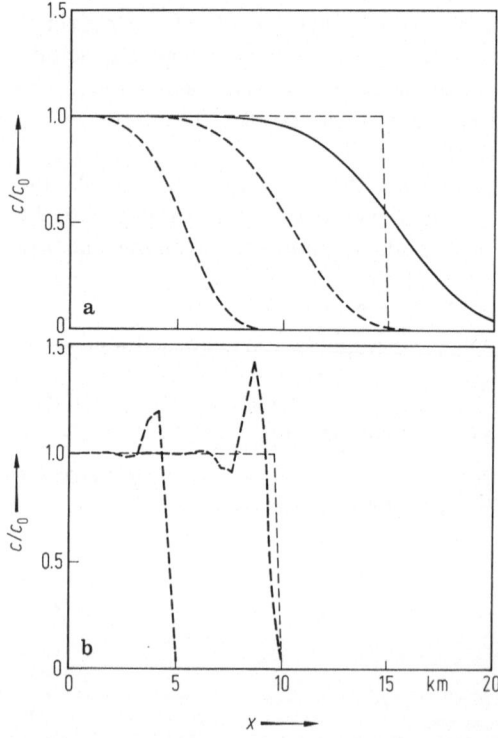

Fig. 5.1 Numerical solutions using the modified Lax method; (a) stable, (b) unstable.

eq. (5.5): the time step cannot be taken arbitrarily large. The physical meaning of this is discussed in section 5.3.

Two examples are shown in Fig. 5.1(a, b). The data are the same as in the example of section 4.2. In the first case, $\alpha = 0.9$ and $\Delta t = 500$ s. Then $\sigma = 0.5$ and the stability condition is satisfied. This is confirmed by Fig. 5.1a, where a smooth solution is seen. Although it is stable, it is not very accurate, as the sharp front is strongly smoothed. In the second example, the time step is increased to 1000 s. Then $\sigma = 1$ and the stability condition is violated. The result is an unstable solution (Fig. 5.1b). In the example of section 4.2 $\alpha = 0$ was used; it is now clear that this is unstable, whatever time step you choose.

5.2 The Leap-Frog Method

You may have noted that time and space derivatives were treated differently: the former by a forward difference and the latter by a central one. Approximating the time derivative by a central difference as well gives the leap-frog method:

$$\frac{c_j^{n+1} - c_j^{n-1}}{2\Delta t} + u\frac{c_{j+1}^n - c_{j-1}^n}{2\Delta x} = 0 \tag{5.6}$$

Its name will be obvious if you consider the pattern of grid points used in this equation. This is a three-level method: if two time levels are known, a third can be computed explicitly (i.e. point by point). However, at $t = 0$, only one initial condition is given, so the leap-frog method cannot be used there. You will have to start the numerical solution using a two-level method (e.g. the modified Lax method) to level $n = 1$. At the downstream boundary, an upstream version of eq. (5.6) can be used.

Considering the stability, assume as before

$$c_j^n = \rho^n \exp{(ij\xi)}$$

Introducing this into eq. (5.6) and dividing by common factors gives a quadratic equation for the amplification factor:

$$\rho^2 + 2\rho\sigma i \sin\xi - 1 = 0 \tag{5.7}$$

with two roots

$$\rho = -\sigma i \sin\xi \pm (1 - \sigma^2 \sin^2\xi)^{1/2} \tag{5.8}$$

It is not difficult to show that these satisfy the Von Neumann criterion if

$$|\sigma| \leqslant 1 \tag{5.9}$$

Even then $|\rho| = 1$ which means that the stability is only marginal: disturbances do not grow, but neither are they damped. Apparently, there are two numerical solutions with the same wave length; one corresponds to the "physical" solution and the other one is a spurious or numerical component. The latter should be kept as small as possible.

The stability condition (5.9) is not too different from that for the modified Lax method; however, the leap-frog method is more accurate (section 5.4)

5.3 The CFL Condition

In the stability analysis you could observe that the dimensionless number

$$\sigma = u\Delta t/\Delta x$$

plays an important part. It is called the *Courant number* and the condition (5.9) the Courant-Friedrichs-Lewy condition after the mathematicians who first noted it. There is a more or less physical explanation, connected to the theory of characteristics discussed in section 4.1. See Fig. 5.2 where the modified Lax method is illustrated; however, the same applies to the leap-frog method.

If you observe the concentration in point x_j, you see the concentration distribution passing by with the characteristic speed u. At time t_{n+1}, the influence of the region below the characteristic through that point is felt; this is called the influence region. Everything above that characteristic is of no importance yet.

In the numerical method, the value at $(j, n+1)$ is determined by the three base points. If $\sigma \leqslant 1$ (Fig. 5.2a), these base points include the influence region and even a little more than that. Apparently, this is not harmful and you get a stable solution.

In the other case, $\sigma \geqslant 1$ (Fig. 5.2b) the base points do not completely include the influence region, so you miss part of the information that is physically important. This cannot be correct and it manifests itself as an instability.

Another way of saying the same.thing is the following. The concentration at $(j, n+1)$ is determined by that in point A. If point A can be interpolated between the grid points (Fig. 5.2a) there is no problem, but extrapolation (Fig. 5.2b) leads to instability.

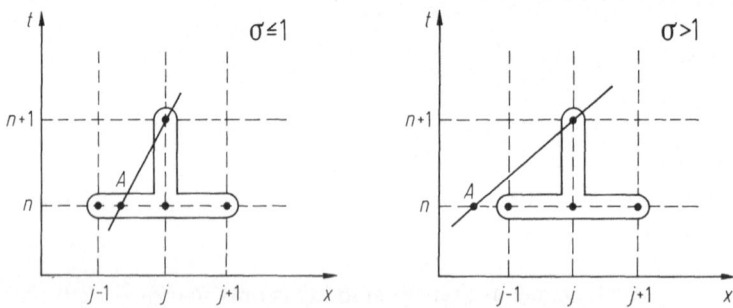

Fig. 5.2 Illustration of the CFL condition.

You should be well aware that the CFL condition (5.9) is necessary but not sufficient. This can be seen in the case of the modified Lax method, where a value of $\alpha < 1$ gives rise to a more restricting condition.

5.4 Truncation Error

A stable finite-difference method is not necessarily an accurate one. There are two ways of investigating the accuracy, one of which is determining the truncation error. If you substitute the solution of the differential equation (which is normally unknown, so it is a mental exercise) into the difference equation, you do not get the differential equation again, but something deviating from it by a (hopefully) small amount. This is seen by developing the solution into Taylor series with respect to a well chosen reference point, normally the point (j, n). This is now a two- dimensional Taylor series:

$$\tilde{c}_j^n = \tfrac{1}{2}\alpha(c_{j+1}^n + c_{j-1}^n) + (1-\alpha)c_j^n = c_j^n + \tfrac{1}{2}\alpha\Delta x^2 \frac{\partial^2 c}{\partial x^2}\Big|_j^n + \ldots$$

$$\frac{c_{j+1}^n - c_{j-1}^n}{2\Delta x} = \frac{\partial c}{\partial x}\Big|_j^n + \frac{1}{6}\Delta x^2 \frac{\partial^3 c}{\partial x^3}\Big|_j^n + \ldots$$

$$\frac{c_j^{n+1} - c_j^n}{\Delta t} = \frac{\partial c}{\partial t}\Big|_j^n + \frac{1}{2}\Delta t \frac{\partial^2 c}{\partial t^2}\Big|_j^n + \ldots$$

Putting this together into eq. (5.1) gives, neglecting second and higher order terms,

$$\frac{\partial c}{\partial t} + u\frac{\partial c}{\partial x} = -\frac{1}{2}\Delta t \frac{\partial^2 c}{\partial t^2} + \frac{1}{2}\alpha\frac{\Delta x^2}{\Delta t}\frac{\partial^2 c}{\partial x^2} + \ldots \tag{5.10}$$

On the left-hand side you recognise the differential equation; the right-hand side should be zero and is called the truncation error. Keep in mind that it is the error in the equation, not in the solution. Equation (5.10) is the one you are actually solving; it is indicated as the *modified equation*. The truncation error is (in this case) proportional to Δt and Δx to the first power; therefore the modified Lax method is of the first order. If Δx and Δt are small enough, the error is small, but how small?

This depends on the time and length scales involved. Suppose you have a travelling wave

$$c = c_0 \sin(kx - \omega t)$$

with $\omega = uk$, then the ratios between the four terms of eq. (5.10) are:

$$\omega : uk : \tfrac{1}{2}\Delta t\,\omega^2 : \tfrac{1}{2}\alpha(k\Delta x)^2/\Delta t$$

To keep the truncation error small, you should have

$$\tfrac{1}{2}\omega\Delta t \ll 1 \tag{5.11}$$

and

$$\tfrac{1}{2}\alpha(k\Delta x)^2/\Delta t \ll uk \tag{5.12}$$

which gives indications on the values of Δx and Δt to be used. However, in practice it is usually difficult to tell what effect a certain truncation error will have on the error in the solution. The latter is the subject of the next section.

As in Chapter 3, there may be more time scales involved, e.g. from a decay term and from external conditions, felt through the initial and boundary conditions. This information should be used in addition to the error analysis given above.

5.5 Wave Propagation

A second way of investigating the accuracy of finite-difference solutions is found by pushing the Von Neumann analysis a little further. In the simple case considered here, the analytic solutions of both the differential and finite-difference equations can be determined and compared with one another. This gives indications that are also valuable in more complicated cases.

As an initial condition we take again

$$c(x, 0) = c_0 \exp(ikx) \tag{5.13}$$

Then the analytic solution for the continuous case is

$$c(x, t) = c_0 \exp\{ik(x - ut)\} \tag{5.14}$$

From the Von Neumann analysis, the finite difference solution is

$$c(x_j, t_n) = c_0 |\rho|^n \exp(ikx + ni\phi) \tag{5.15}$$

where ϕ is the argument of the amplification factor ρ. The ratio of the amplitudes in eqs. (5.14) and (5.15) after n steps is the damping factor

$$d(n) = |\rho|^n \tag{5.16}$$

and the ratio of the phase angles, or the velocities of propagation

$$c_r = -n\phi/ukt = -\phi/uk\Delta t = -\phi/2\pi\sigma\xi \tag{5.17}$$

The quantity c_r is called the relative velocity of propagation. Note that these expressions are valid for any finite-difference method; only by introducing the specific expression for ρ, you get the values for a particular method. Both factors d and c_r should be near unity to have an accurate solution. By adjusting σ and ξ, i.e. Δx and Δt, you can influence the error. Contrary to the stability analysis, where all possible wave lengths are involved, the present analysis needs to be done only for those wave lengths you are interested in, i.e. that are physically important.

The question what errors are acceptable is not an easy one to answer. It depends very much on the particular application. An example is given in Chapter 6.

5.6 Exercises

1. Show that eq. (5.1) is a general approximation of the simple wave equation. To do this, write the finite-difference equation as

$$c_j = pc_{j-1} + qc_j + rc_{j+1}$$

Develop this into Taylor series and equate the coefficients with the corresponding ones in the differential equation. You will find that there are two conditions for the three coefficients p, q, r, so one degree of freedom is left, which can be formulated in terms of α as in eq. (5.1).
2. Determine the truncation error of the leap-frog method and show that it is of the second order, so more accurate than the modified Lax method.
3. Determine the truncation error of the upstream method eq. (4.11).

Chapter 6
Kinematic Waves

6.1 Theory

In the theory of long waves in rivers, to be discussed in chapter 15, an extreme case can be considered in which the momentum equation reduces to a simple equilibrium between bottom friction and gravity (for a more detailed discussion see that chapter):

$$c_f \frac{Q|Q|}{RA_s^2} = gi \tag{6.1}$$

where c_f = bottom friction coefficient

A_s = area of cross-section

R = hydraulic radius

i = bottom slope

The mass balance for a "slice" of the river is eq. (4.2) or

$$B \frac{\partial a}{\partial t} + \frac{\partial Q}{\partial x} = 0 \tag{6.2}$$

where

B = surface (or storage) width

a = water depth

Q = discharge

As A_s and R in eq. (6.1) are functions of the water depth, you have now two equations with the unknowns Q and a. It is quite simple to eliminate Q by solving it from eq. (6.1) and substituting it into eq. (6.2):

$$\frac{\partial a}{\partial t} + c \frac{\partial a}{\partial x} = 0 \tag{6.3}$$

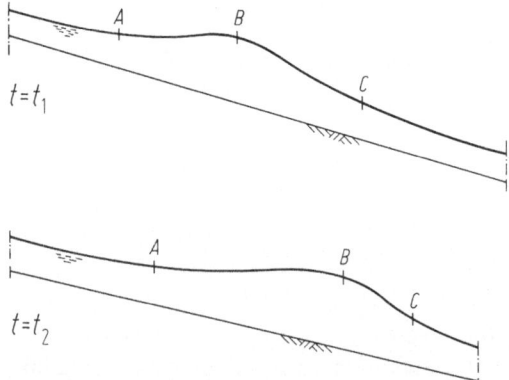

Fig. 6.1 Deformation of a kinematic wave.

where

$$c = B^{-1}(gi/c_f)^{1/2}\frac{\mathrm{d}}{\mathrm{d}a}(R^{1/2}A_s)\tag{6.4}$$

Equation (6.3) is in exactly the same form as the simple wave equation (4.3) without any decay term. The velocity of propagation is now c, given by eq. (6.4). Along characteristics moving with this velocity, the water depth remains constant. This is true in particular for the wave crest. Therefore, in this kinematic-wave approximation, a flood wave in a river does not decay.

To get an idea on the velocity of propagation, consider a river with a wide, rectangular cross-section, where $A_s = B_s a$ and B_s is the width of the cross-section inasmuch as it carries the flow. This is not necessarily equal to the storage width B, as there may be "dead zones" in the river which do not contribute to the flow, but which do serve as storage basins. If the width-to-depth ratio is large, the hydraulic radius R is approximately equal to the depth. Introducing all this into eq. (6.4), you get

$$c = \tfrac{3}{2}u B_s/B\tag{6.5}$$

which is of the same order of magnitude as the flow velocity. Having a great part of storage area not contributing to the flow ($B_s/B < 1$) results in a slow-down of the wave. Another interesting point is that c is not a constant: it depends on the water depth and the wave crest travels at a greater speed than the wave trough (Fig. 6.1). Still, the crest level remains constant, but the front of the wave gets steeper.

6.2 Example

To protect a city on a river from flooding, the construction of a storage basin 400 km upstream is considered. The effect of this is studied by using a mathematical model. The data are as follows:

(i) Dimensions of the river
For simplicity assume constant width and slope:

$$B_s = B = 200 \text{ m}$$
$$c_f = 0.004$$
$$i = 10^{-4}$$

(ii) Initial condition
This is not known in principle, so you have to make a reasonable assumption. Suppose this is an equilibrium flow of $Q = 800 \text{ m}^3/\text{s}$, corresponding to an equilibrium depth of 4 m. If you take a non-equilibrium flow as initial condition, waves will be generated that travel along the river.

(iii) Boundary condition
On the upstream side, a boundary condition is needed, that could be either a record of the waterlevel far upstream or the discharge hydrograph for a design flood, based on a hydrologic analysis of the river catchment area. The discharge can be translated into a water level using eq. (6.1). In this case assume a single sinusoidal wave with height 4 m above equilibrium and a period of $T = 6$ days (Fig. 6.2). The boundary is located at 450 km upstream of the city.

(iv) Dimensions of the basin
The basin is schematized as a concentrated storage area with a surface of $5 \ 10^8 \text{ m}^2$, that starts being filled as soon as the local waterlevel exceeds 6.40 m.

The numerical schematization is based on the modified Lax method. For the example, a modest accuracy is chosen: the waterlevel at the city should be reproduced with an error of less than 25 cm. Choose, e.g. $\alpha = 0.8$ and $\sigma = 0.8$. This satisfies the stability condition. The maximal value $\sigma = 1$ can hardly be used here as the speed of propagation, and therefore the Courant number, varies with the water depth. Try some values of the grid size and compute the numerical damping and relative velocity of propagation:

$L/\Delta x$	$T/\Delta t$	n	$d(n)$	c_r
100	125	50	0.97	1.0002
50	62	25	0.94	1.0009
25	31	12.5	0.88	1.004

With the velocity of propagation of a kinematic wave, the wave length L is approximately 1000 km. The travel time over 450 km is 60 h or 0.4 T so the number of time steps required is $n = 0.4 \ T/\Delta t$. It turns out that for $L/\Delta x = 50$ to 100 the required accuracy is met. Note that the value of c_r is of no interest here as no conditions of accuracy for the time of arrival of the wave have been given; it is however quite close to unity. This means that $\Delta x = 10$ to 20 km and correspond-

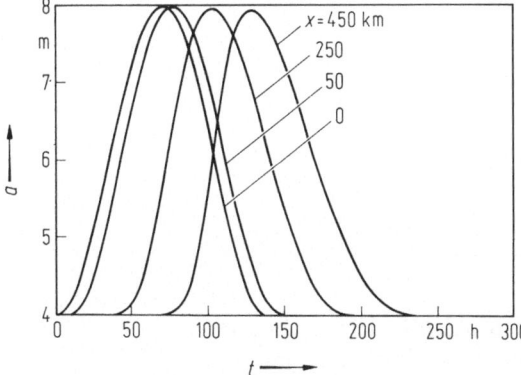

Fig. 6.2 Existing situation; water level as a function of time at four locations.

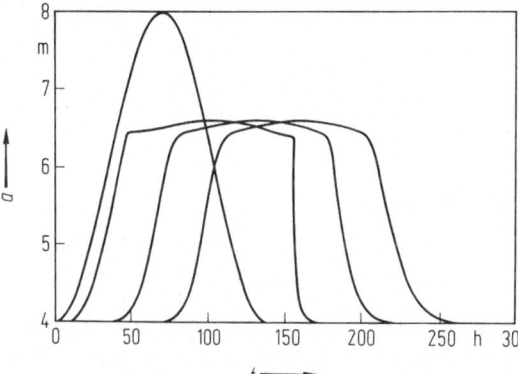

Fig. 6.3 Situation with storage basin; same locations.

ingly (from the Courant number) $\Delta t = 1$ to 2 h. The smaller values have been used for the numerical example.

These values are rather small considering the length and duration of the wave. The reason is that the modified Lax method is rather inaccurate, particularly for the wave damping. You can check for yourself what happens if you use the leap-frog method.

The result for the case without storage basin is shown in Fig. 6.2. The properties of a kinematic wave can be clearly seen. The small amount of damping is due to the numerical method. It should be noted that the wave in reality would reduce in amplitude some 0.5 m over this distance. Therefore, a much higher numerical accuracy would not be very useful. Figure 6.3 gives the result with the projected storage basin. It turns out to work quite effectively: the waterlevel hardly exceeds 6.40 m.

Chapter 7
Diffusion

7.1 Groundwater Flow in a Horizontal Layer

Consider the flow of groundwater in a more or less horizontal porous layer of soil (Fig. 7.1). The mass balance for a column of soil is similar to that for a river:

$$\frac{\partial a}{\partial t} + \frac{\partial}{\partial x}(au) = w/n \tag{7.1}$$

where

a = thickness of the water layer
u = flow velocity averaged over the layer
w = rainfall in volume per unit surface area
n = porosity of the soil

It has been assumed that rainfall reaches the groundwater layer very quickly; temporary storage in the unsaturated soil above the groundwater layer is not taken into account.

As a dynamic equation for groundwater flow the Dupuit relation is often used (see Verruijt 1970 for a more complete discussion):

$$u = -k\frac{\partial h}{\partial x} \tag{7.2}$$

in which

h = surface level of groundwater relative to a horizontal plane
k = constant of proportionality expressing the permeability of the soil (it is a soil property with the dimension m/s)

Combining the two equations

$$\frac{\partial h}{\partial t} - \frac{\partial}{\partial x}\left(ka\frac{\partial h}{\partial x}\right) = w/n \tag{7.3}$$

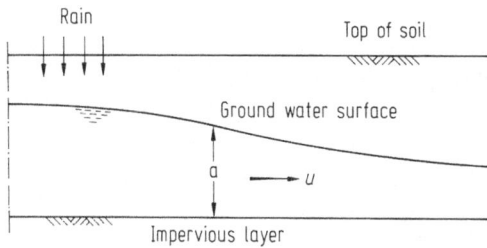

Fig. 7.1 Nearly horizontal ground-water flow.

which is the basic equation for this chapter. In many cases, it can be assumed that the depth of the layer a does not vary too much and the combination $D = ka$ is approximately constant (moreover the permeability k is often not very well known so it may not be very useful to take small variations in the layer thickness into account). Then you get the standard diffusion equation

$$\frac{\partial h}{\partial t} - D \frac{\partial^2 h}{\partial x^2} = w/n \tag{7.4}$$

in which D is the diffusion coefficient (dimension m^2/s). This type of equation again occurs quite often in physics, e.g. in heat conduction. It is a differential equation of the first order in t and second order in x. Its properties are basically different from the simple-wave equation discussed before.

You get an idea of the phenomenon of diffusion by considering the example of a groundwater layer next to a river. Suppose everything is in equilibrium (no rainfall, groundwater surface horizontal). For some reason the waterlevel in the river is suddenly lowered (i.e. quickly, compared with the response of the groundwater) and then remains constant again. Groundwater will now flow towards the river as there is a surface slope (eq. 7.2) as shown in Fig. 7.2.

Unfortunately, the solution of eq. (7.4) cannot be found with elementary means (unless you are familiar with Laplace transforms), but you can check by substitution that it is

$$h(x, t) = h_0 - \Delta h\, erfc\left(\frac{x}{2\sqrt{Dt}}\right) \tag{7.5}$$

in which $erfc$ is a standard mathematical function (the "complementary error

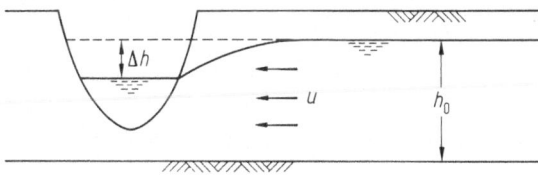

Fig. 7.2 Groundwater flow due to fall of waterlevel in river.

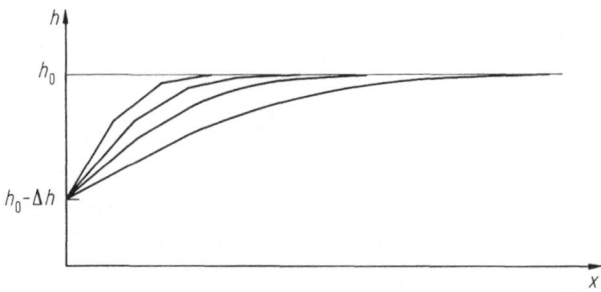

Fig. 7.3 Solution of diffusion equation for situation of Fig. 7.2.

function") defined as

$$erfc(z) = \frac{2}{\sqrt{\pi}} \int_z^\infty e^{-t^2}\, dt \tag{7.6}$$

It goes to zero for $z \to \infty$ and it can be shown to be 1 for $z = 0$. This solution is shown in Fig. 7.3 for some instants of time. From eq. (7.5) and Fig. 7.3, some typical properties of diffusion can be concluded.

(i) After any time t, however small it may be, the influence of the disturbance at $x = 0$ is felt everywhere (though perhaps only very weakly). The disturbance therefore extends itself at an infinite speed.

(ii) The decrease of the groundwater level is the same for $x/t^{1/2} =$ const. so the expansion of the disturbance is not linear with time but behaves like $t^{1/2}$. The expansion is also proportional with $D^{1/2}$. A significant influence is felt at a value of (roughly) $x = (4Dt)^{1/2}$.

(iii) At a distance L from the river, you will in principle notice the change immediately, but a real change of groundwater level will be observed only after a time of the order $L^2/4D$.

The need for initial and boundary conditions is a little more difficult to show than for the simple-wave equation, but you will probably find the following rules plausible (for more detail see any good book on partial differential equations, such as Garabedian, 1967). As the diffusion equation is of first order in time, you need one initial condition, usually the groundwater level at $t = 0$. You do not have to specify the other variable u, as it can be derived from h using eq. (7.2).

The equation is of second order in space, so you need two boundary conditions: one on each side. In the example these are:

– the groundwater level at the river ($x = 0$) which is supposed to be equal to the river waterlevel;

– the groundwater level at infinity (or at least far from the river) which is supposed to remain constant.

7.2 Explicit Finite-Difference Method

If you want to solve the diffusion equation numerically, you have to find an approximation for the second derivative. This can be most easily done by repeating the process for a first derivative:

$$\frac{\partial^2 h}{\partial x^2} \approx \frac{1}{\Delta x}\left(\left.\frac{\partial h}{\partial x}\right|_j - \left.\frac{\partial h}{\partial x}\right|_{j-1}\right) \approx \frac{1}{\Delta x^2}(h_{j+1} - h_j - h_j + h_{j-1})$$

This gives a central finite-difference approximation. Combining this with a forward time difference, you get the forward time, central space method or FTCS method:

$$\frac{h_j^{n+1} - h_j^n}{\Delta t} - D\frac{h_{j+1}^n - 2h_j^n + h_{j-1}^n}{\Delta x^2} = w/n \tag{7.7}$$

This can be started from the given initial condition. If, at a boundary, h is known from a boundary condition, no special approximations are needed there. However, if $\partial h/\partial x$ is given, this has to be discretized, e.g. as

$$(h_1^{n+1} - h_0^{n+1})/\Delta x = g^{n+1}$$

By now, you have been warned against instability, so you will check for that. The amplification factor is easily found to be

$$\rho = 1 + \lambda(\cos \xi - 1) \tag{7.8}$$

in which a new dimensionless parameter $\lambda = 2D\Delta t/\Delta x^2$ occurs, that, for lack of a better name, is called the diffusion parameter. As the amplification factor in eq. (7.8) is real, it is easy to show that it does not exceed unity in absolute value for any ξ if

$$0 \leqslant \lambda \leqslant 1 \tag{7.9}$$

Here, again, you meet a time-step restriction which is now proportional to the grid size squared. This is particularly bad if you want to use a small grid size for accuracy.

As an example, take the following data

$$D = 10^{-3} \text{ m}^2/\text{s}$$
$$\Delta x = 10 \text{ m}$$
$$\Delta t = 15 \text{ h}$$

The value of the waterlevel decrease Δh is not important as the diffusion equation is linear. Figure 7.4a shows the computed solution after 10, 20 and 30 time steps. Apparently, this case is unstable, which is confirmed by the value of $\lambda = 1.08$. If you halve the time step, the solution of Fig. 7.4b is found which is stable.

The diffusion parameter λ in a sense is comparable to the Courant number. It was just mentioned that the influence of a disturbance by diffusion over a length L takes a time of the order $L^2/4D$. For one grid interval Δx, this will be $\Delta x^2/4D$. The condition of stability (7.9) expresses that the time step should not exceed this

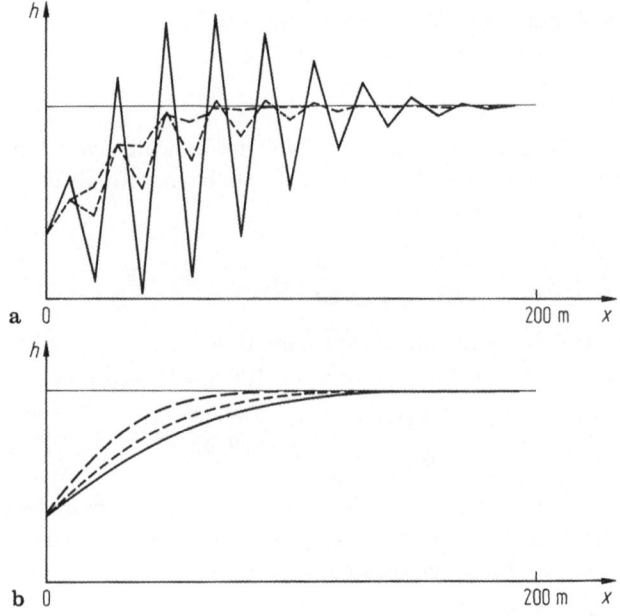

Fig. 7.4 Decrease of groundwater level with FTCS method: (a) unstable, (b) stable.

"diffusion time". That there is a slightly different constant involved is not essential as it depends on what you call a "significant" influence.

7.3 Implicit Finite-Difference Method

There is a way to circumvent stability restrictions such as eq. (7.9) by using implicit methods. The idea has already been put forward in section 3.4. The principle is to approximate the differential equation not at time t but at some intermediate time $t + \theta \Delta t$ with $0 \leqslant \theta \leqslant 1$. Then you still have

$$\frac{\partial h}{\partial t}\bigg|_{t_n + \theta \Delta t} \approx \frac{h_j^{n+1} - h_j^n}{\Delta t} \tag{7.10}$$

but for the spatial derivative you have to take an average between the two time levels, as in the discrete grid you do not have any intermediate values:

$$\frac{\partial^2 h}{\partial x^2} \approx \theta \frac{h_{j+1}^{n+1} - 2h_j^{n+1} + h_{j-1}^{n+1}}{\Delta x^2} + (1 - \theta) \frac{h_{j+1}^n - 2h_j^n + h_{j-1}^n}{\Delta x^2} \tag{7.11}$$

Putting these together into the diffusion equation results in the *Crank–Nicholson* method:

$$h_j^{n+1} - h_j^n = \lambda\{\theta(h_{j+1}^{n+1} - 2h_j^{n+1} + h_{j-1}^{n+1}) +$$
$$+ (1 - \theta)(h_{j+1}^n - 2h_j^n + h_{j-1}^n)\} \tag{7.12}$$

The difficulty is that this equation contains three unknown values at time level $n + 1$, so it cannot be solved by itself. However, if you count the equations, you will see that you have exactly the right number: one of eq. (7.12) for each internal (i.e. non-boundary) point, and one boundary condition at each of the two boundary points. Therefore, all unknowns can be solved from the system. It is no longer possible to find one value explicitly from the equations, but all of them are well-defined implicitly, which explains the name of this type of methods. The same idea can be applied to the simple-wave equation. An elegant way of solving the system of equations is described in section 7.4. An example of an implicit solution is given in Fig. 7.5 which concerns the same situation as Fig. 7.4. Here, $\theta = \frac{1}{2}$ and $\Delta t = 28$ h or 140 h, corresponding to $\lambda = 2$ or 10 respectively. Both solutions are perfectly stable, though apparently not equally accurate.

That this method is indeed stable for values of $\lambda \geqslant 1$ can be shown using the amplification factor, which now reads (please check)

$$\rho = \frac{1 + (1 - \theta)\lambda(\cos\xi - 1)}{1 - \theta\lambda(\cos\xi - 1)} \tag{7.13}$$

Using this, it is simple to show that the method is stable for any value of λ if

$$\tfrac{1}{2} \leqslant \theta \leqslant 1$$

You may wonder whether this agrees with the explanation in section 7.2 on the significance of the diffusion parameter. In this case, the unknown value at a point j is influenced by its neighbours, those in turn by theirs, and so on to the boundary points. Also, all values of the old time level have their influence. Therefore, a disturbance can have its influence over the entire region during one time step, in agreement with the properties of the differential equation. This is true for any value of the time step.

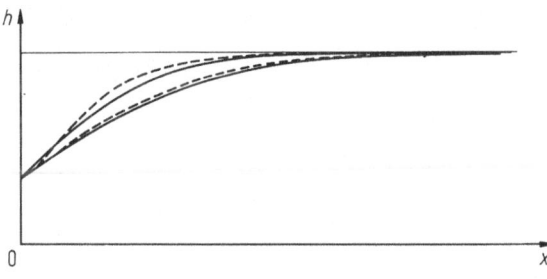

h

0

x

Fig. 7.5 Implicit solution after 280 and 560 h; $\lambda = 2$ (drawn lines) or 10 (dashed lines).

7.4 The Thomas Algorithm

The price paid for unconditional stability of the Crank–Nicholson method is that you have to solve a system of algebraic equations at every time step. This is, however, a system with a very special form that makes it easy to solve. Together with the boundary conditions, the system can be written as

$$
\begin{bmatrix}
1 & 0 & 0 & 0 & . & & . \\
-\frac{1}{2}\theta\lambda & 1+\lambda\theta & -\frac{1}{2}\theta\lambda & 0 & . & & . \\
0 & -\frac{1}{2}\theta\lambda & 1+\lambda\theta & -\frac{1}{2}\theta\lambda & . & & . \\
0 & 0 & . & . & . & & . \\
. & . & 0 & -\frac{1}{2}\theta\lambda & 1+\lambda\theta & -\frac{1}{2}\theta\lambda \\
. & . & . & 0 & 0 & 1
\end{bmatrix}
\times
$$

$$
\times
\begin{bmatrix}
h_0^{n+1} \\
h_1^{n+1} \\
\vdots \\
h_{J-1}^{n+1} \\
h_J^{n+1}
\end{bmatrix}
=
\begin{bmatrix}
f_0^{n+1} \\
b_1 \\
\vdots \\
b_{J-1} \\
f_J^{n+1}
\end{bmatrix}
\qquad (7.14)
$$

with $b_j = h_j^n + \frac{1}{2}(1-\theta)\lambda(h_{j+1}^n - 2h_j^n + h_{j-1}^n)$.

The matrix of coefficients contains a great deal of zeros, except at the main diagonal and its two neighbours. It is called a tridiagonal matrix.

The solution of such a system can be done very effectively using the *Thomas* or double sweep algorithm, which is in fact nothing else than the standard Gauss elimination method taking the zeros into account. If you write any of the equations in the form

$$a_j h_{j-1} + b_j h_j + c_j h_{j+1} = d_j \qquad (7.15)$$

with $a_0 = c_J = 0$, it is possible to state

$$h_j = e_j + f_j h_{j+1} \qquad (7.16)$$

provided

$$e_j = \frac{d_j - a_j e_{j-1}}{b_j + a_j f_{j-1}} \qquad e_0 = \frac{d_0}{b_0} \qquad (7.17)$$

$$f_j = \frac{-c_j}{b_j + a_j f_{j-1}} \qquad f_0 = -\frac{c_0}{b_0} \qquad (7.18)$$

You can verify this by substituting eq. (7.16) for $j - 1$ into eq. (7.15). First the equations (7.17) and (7.18) are solved (in the forward sweep) for $j = 1, 2, \ldots, J$. Then h_J is known from the boundary condition. Finally, in the backward sweep the other unknowns are solved from eq. (7.16) for $j = J - 1, J - 2, \ldots, 1$. The process is efficient both in computer time and storage.

7.5 Application

In a drained agricultural area, there are parallel drainage channels at a distance L of one another. During a time T it is raining with an intensity $q = w/n$. What is the highest groundwater level reached and how much time does it take to return to the original situation?

Suppose that the soil is homogeneous; then the situation between the channels is symmetric and you need only consider an area of width $L/2$. At the line of symmetry, the boundary condition is then $\partial h/\partial x = 0$. In the drainage channels, suppose that the waterlevel remains constant: $h = 0$. This same waterlevel is taken as the initial condition.

With $L = 100$ m, take 20 grid intervals of $\Delta x = 2.5$ m for the half region. The diffusion coefficient is, as in the preceding example, $D = 10^{-3}$ m^2/s. The duration of the rain period is $T = 29$ days. To reproduce this, a time step of $\Delta t = 0.5$ days seems to be small enough. Then $\lambda = 15$, so an implicit method is needed. Using the Crank–Nicholson method with $\theta = \frac{1}{2}$, Fig. 7.6 results, which gives the dimensionless increase of groundwater level. About an interval T after the end of the rainfall period the level is back to normal.

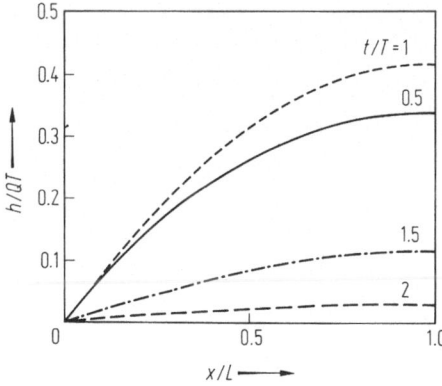

Fig. 7.6 Increase of groundwater level during rainy period.

7.6 Exercises

1. In a large plain, you have a layer of groundwater of approximately 20 m of depth. The permeability coefficient is 10^{-4} m/s. You want to extract part of the water for irrigation or domestic purposes. To this end, a line of wells is constructed; consider this as a distributed withdrawal of q m^2/s. You want to simulate the behaviour of the groundwater layer numerically, taking into account that there is a rainfall period of 3 months. In a cross section normal to the line of wells, you can use the one-dimensional Dupuit formulation discussed in this chapter.

   (i) Where would you choose the boundaries of the computational region?

   (ii) What boundary conditions would be used?

   (iii) Estimate the grid size Δx.

   (iv) Estimate the time step Δt.

   (v) Would you use an explicit or an implicit method?

2. It could be considered to use central instead of forward time differences in the numerical solution of the diffusion equation. This would result in a sort of leap-frog method (compare to eq. 5.6). Show that this is useless, as the resulting method is unconditionally unstable (i.e. it is not possible to find a time step for which it is stable).

Chapter 8
Numerical Accuracy for Diffusion Problems

8.1 Fourier Series

To study the numerical accuracy for diffusion problems, it is no longer sufficient to consider one sinusoidal wave, as in previous chapters. You must take into account that an arbitrary function, acting as initial condition, can be thought to be built up from a series of sinusoidal functions, which is called a Fourier series. As an example take a block function with length $2L$ (Fig. 8.1). If you consider this on a larger interval of arbitrary length $2A$, it can be developed into a Fourier series as:

$$h(x, 0) = a_0 + \sum_{j=1}^{\infty} a_j \cos (2\pi x/L_j) \tag{8.1}$$

where $L_j = 2A/j$ is the wave length of component j, and the amplitudes are determined from

$$a_j = \frac{1}{2A} \int_{-A}^{+A} h(x, 0) \cos (2\pi x/L_j) dx \tag{8.2}$$

$$a_j = \frac{1}{2\pi} L_j/A \sin (2\pi A/L_j) \tag{8.3}$$

Due to the symmetric form of the· block function, the sine terms are zero. The "strength" of the Fourier components, expressed by the amplitudes a_j, is illustrated in Fig. 8.2 as a function of the wave length. This is sometimes called the amplitude spectrum.

This figure shows that the principal contributions are those with wave lengths greater than (roughly) $2L$ which is the length of the block. Components with smaller wave lengths are present, but not very important. If you do this for a another type of function with the same length $2L$, you get a slightly different amplitude spectrum, but you will draw the same conclusion on the principal wave lengths (see the exercises for this chapter). This can therefore be used as a rule of thumb for an arbitrary initial function. Note that the length of the interval $2A$ indeed does not enter in this estimate.

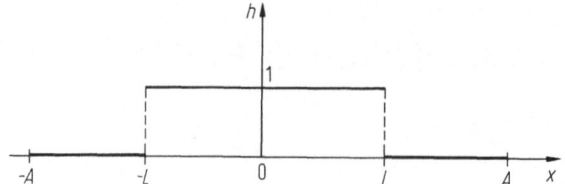

Fig. 8.1 Block function.

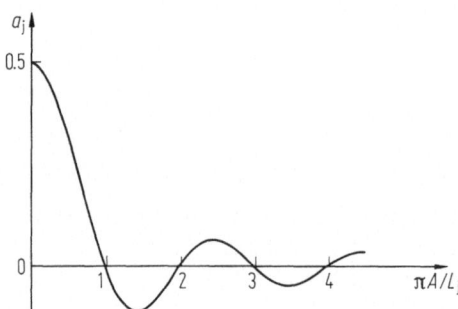

Fig. 8.2 Fourier components of block function.

8.2 Transfer Function

Once you have determined which Fourier components in the initial condition are most important, the next question is what happens to them in the diffusion equation. Take an arbitrary term as initial condition:

$$h(x, 0) = a \cos (kx) \tag{8.4}$$

with wave number $k = 2\pi/L_j$, then it is easily checked that the solution of the diffusion equation is

$$h(x, t) = H a \cos (kx) \tag{8.5}$$

with a "transfer function" (Fig. 8.3)

$$H(k) = \exp (-Dk^2 t) \tag{8.6}$$

This indicates that each wave component is multiplied by a damping factor that, for a fixed time, depends on the wave length (and of course on the diffusion coefficient), such that shorter waves are damped more strongly than longer waves. So, as time progresses, more and more of the short-wave contributions that were originally present get damped out. This agrees with the general character of the diffusion equation that steep gradients are smoothed (Fig. 7.2).

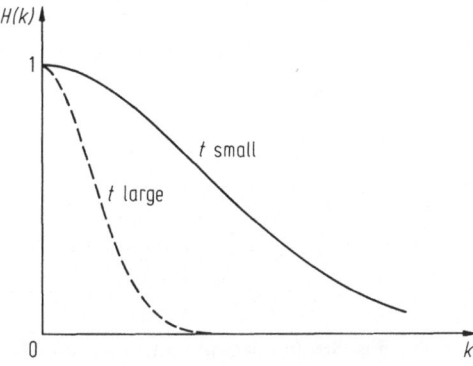

Fig. 8.3 Transfer function for a diffusion equation.

8.3 Numerical Representation

Those Fourier components that are important in the intial condition, and that still are important at the time of interest t, should be represented with a specified numerical accuracy. In other words: the numerical transfer function should be close to the analytical one for the important wave lengths. There is no use in trying to represent short waves accurately, if they are damped out by the system anyway.

Starting from the same initial condition (8.4), the numerical solution is

$$h(x, t) = \rho^n a \cos (kx) \tag{8.7}$$

where $n = t/\Delta t$ is the number of time steps and ρ is the amplification factor of the numerical method used. The numerical transfer function is therefore

$$H_n = \rho^n \tag{8.8}$$

This can be compared with the analytical transfer function from eq. (8.6) and Δt and Δx can be determined such that the two agree within some required accuracy. Some examples are given in Chapters 9 and 10.

The result depends very much on the total time t of the simulation. As an illustration, consider the time $t = (k^2 D)^{-1}$ in which the amplitude reduces to e^{-1} times its original value (relaxation time; note that this is different for each component). Then

$$n = t/\Delta t = (k^2 D \Delta t)^{-1} \tag{8.9}$$

If the grid interval is not too large (such that $\xi < 1$), a reasonable approximation for the amplification factor of the Crank–Nicholson method is:

$$\rho \approx -\frac{1 - \frac{1}{2}(1 - \theta)\lambda\xi^2}{1 + \frac{1}{2}\theta\lambda\xi^2} \tag{8.10}$$

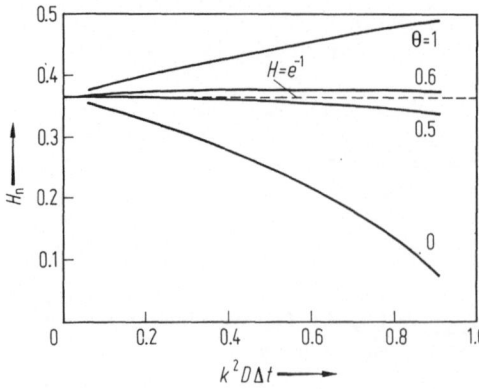

Fig. 8.4 Numerical transfer function for $k^2 Dt = 1$.

in which $\frac{1}{2}\lambda\xi^2 = k^2 D\Delta t$. This is apparently the relevant factor determining the numerical accuracy. In Fig. 8.4, the numerical transfer function is shown as a function of this parameter. The analytical value e^{-1} is also indicated. A few conclusions can be drawn:

(i) the factor θ has a very great influence. Values of 0.5 or close to that are clearly preferable;

(ii) the time step should be less than 10% of the relaxation time to have any accuracy at all, if $\theta = 0$ or 1. For $\theta = 0.5$, a time step 5 times greater is acceptable;

(iii) for $\theta = 1$, the accuracy is about as bad as for $\theta = 0$. Similar values of the time step result for the two cases. The difference is that you do not have to bother about stability if $\theta \geqslant 0.5$.

8.4 Exercises

1. Determine the Fourier components of the function

$$h(x, 0) = \cos(\pi x/2L) \quad \text{if } |x| \leqslant L$$
$$= 0 \qquad\qquad \text{if } |x| > L$$

Show that the principal components have wave lengths of $2L$ or greater.

2. Compute the transfer function (both the analytical and the numerical one) for the example of section 7.5. Which are the relevant wave lengths in that case?

Chapter 9
Diffusion Model for Coastline Development

9.1 Mathematical Formulation

For the global behaviour of a coastline there exists a simple theory which, without going deeply into the physics, can give a useful insight in the consequences of breakwater construction, beach nourishment etc. The approach is originally due to Pelnard-Considère.

If waves approach the coastline obliquely, a wave-driven flow along the coast is generated. On a sandy coast, the flow together with the stirring action of breaking waves causes a longshore sand transport. The magnitude of this is not easily determined, but an estimate can be made using (e.g.) the CERC formula

$$s = E \sin 2\phi \tag{9.1}$$

where

s = volume of sand transported along the coast (m³/s)

ϕ = angle of wave incidence outside the breaker zone (Fig. 9.1)

E = coefficient proportional to the incident wave energy flux (m³/s)

In eq. (9.1) the angle 2ϕ occurs, which can be made plausible as follows. If the waves run into the beach at right angles ($\phi = 0$), the longshore transport will be zero. On the other hand, if the waves propagate parallel to the coast ($\phi = \pi/2$), they do not reach the shore, so there will be no sand transport either. In between these extremes, there will be some maximum; eq. (9.1) at least qualitatively represents this behaviour.

If the angle of incidence ϕ and/or the wave height (and consequently the coefficient E) vary, the longshore transport varies from one section to the other. As a consequence, sand is deposited or eroded and the coastline advances or recedes. It is assumed that there is no lateral (on/offshore) sand transport. Then a sand balance equation can be set up for a "slice" of the coastal region (Fig. 9.2). An additional assumption is that the shape of the beach profile is unchanged, so that the coastline as a whole moves forwards or backwards down to a depth a below

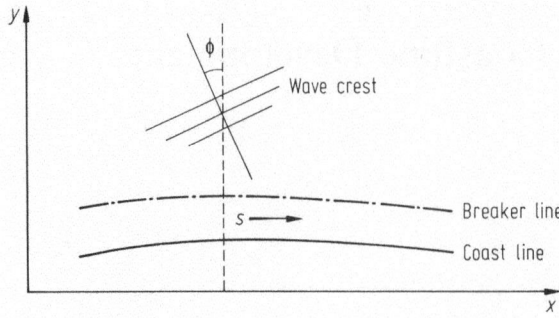

Fig. 9.1 Wave incidence and longshore sand transport.

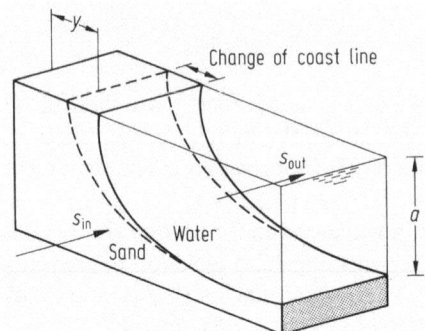

Fig. 9.2 Sand balance for a coastal section.

which there is no sand transport anymore

$$a\frac{\partial y}{\partial t} + \frac{\partial s}{\partial x} = 0 \tag{9.2}$$

where y is the distance of the shoreline to a fixed straight line of reference. If ϕ_0 is the angle of wave incidence relative to this reference frame, the angle between the wave fronts and the local orientation of the coastline is

$$\phi = \phi_0 - \frac{\partial y}{\partial x} \tag{9.3}$$

Putting eqs. (9.1), and (9.3) into (9.2) gives an equation for the coastline position y:

$$a\frac{\partial y}{\partial t} + 2\phi\frac{\partial E}{\partial x} + 2E\frac{\partial \phi_0}{\partial x} - 2E\frac{\partial^2 y}{\partial x^2} = 0 \tag{9.4}$$

Here, we consider only those cases where E and ϕ_0 are constants; then a simple diffusion equation results:

$$\frac{\partial y}{\partial t} - D\frac{\partial^2 y}{\partial x^2} = 0 \tag{9.5}$$

with a diffusion coefficient

$$D = 2E/a \tag{9.6}$$

An estimate of this can often be made as follows. At the undisturbed coastline, an equilibrium sand transport s_0 is supposed to be known from field measurements. Then from eq. (9.1) and (9.6), E can be eliminated (which is a quantity particularly difficult to determine), so you get

$$D = s_0(\phi_0 a)^{-1} \tag{9.7}$$

9.2 Initial and Boundary Conditions

As an initial condition, you need the position $y(x, 0)$ of the coastline over the entire reach of interest at a time $t = 0$. This may (but need not) be an equilibrium situation.

According to the rule for diffusion equations, you need a boundary condition at each boundary on both sides of the reach. There are a few possibilities:

- far away $(x \to \infty)$ there is no change, so either $y = 0$ or $\partial y/\partial x = 0$.
- at a construction such as a breakwater the sand flux is zero; as the wave height is not, you have to conclude from eqs. (9.1) and (9.3) that $\partial y/\partial x = \phi_0$. This means that the coastline adjusts itself such that it is parallel to the incident wave crests near a fixed construction (Fig. 9.3);
- discharge of sand from a river; this constitutes an "internal boundary" at which $s^+ - s^- = \Delta s$ where the three terms represent the longshore transport on both sides of the discharge point and the sand discharge from the river. Using eqs. (9.1) and (9.3) this gives a relation involving the coastline position y.

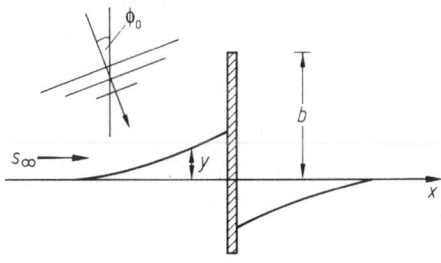

Fig. 9.3 Situation near breakwater.

9.3 Example

Given is a straight coastline in equilibrium, with data as in exercise 1. Then you can conclude $D = 0.42 \ 10^6 \ \mathrm{m^2/year}$.

A breakwater of length $b = 300$ m is considered, which for a while intercepts the complete longshore sand transport.

The initial condition is, of course, a straight coastline. Boundary conditions far from the breakwater are such that no disturbances occur, i.e. $y = 0$. At the breakwater, you have an internal boundary where on both sides the sand flux is zero, i.e. $\partial y / \partial x = \phi_0$. However, the values of y will be different on both sides of the breakwater. Actually, if no sand passes the breakwater, both sides of the region are completely decoupled.

Sand will accumulate on the upstream side of the breakwater. After some time, the available space will have been filled up and sand will start passing the tip of the breakwater. From that instant of time, a different boundary condition is needed: you can then assume that the coastline does not advance any further, so $y = b$ on the upstream side. The surplus of sand passes the breakwater and is transported on the downstream side. Due to continuity of sand transport, you will now have

$$\frac{\partial y}{\partial x} \text{ (upstream)} = \frac{\partial y}{\partial x} \text{ (downstream)} \tag{9.8}$$

In order to determine the grid size and time step for a numerical model, it is useful to have some idea of the time and length scales involved. A very rough estimate is sufficient. For example, you could assume that all the intercepted sand accumulates in a triangle upstream of the breakwater, with the coastline parallel to the wave crests (angle ϕ_0). The volume of that is $4 \ 10^6 \ \mathrm{m^3}$ and the length about 1500 m. Given the equilibrium sand transport, it will take some 3 years to fill up the triangle.

Using the theory of Chapter 8, you see that a local disturbance (with zero length) is introduced at $t = 0$. Therefore, initially, all wave lengths up from zero are present. Secondly, the transfer function determines which wave lengths play any part after a typical time which, from the preceding estimate, may be taken 3 years. Disregarding everything that is damped to less than (say) 10%, you have

$$\exp\left(-k^2 Dt\right) > 0.1$$

or $k^2 Dt < 2.3$. With $t = 3$ years, this gives $k < 1.4 \ 10^{-3}$ or wave length $L > 5$ km approximately. To reproduce this wave length, a grid size of (say) 0.25 km is good enough.

Thirdly, for the time step, you require that the numerical transfer function for $t = 3$ years and the given wave length does not deviate more than (say) 5% from the analytical value, which was chosen as 0.1. With some experimentation (please check) you find that $\Delta t = 0.06$ years satisfies this, using an explicit method ($\theta = 0$). You should check the stability condition, giving $\lambda = 0.84$, which is satisfactory.

Using these data, the results of the numerical model are those shown in Fig. 9.4

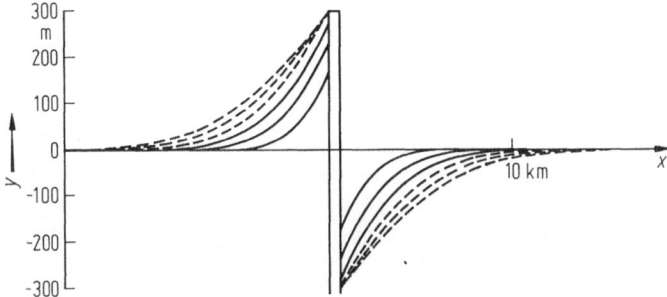

Fig. 9.4 Computed coastline at one-year intervals.

at one-year time intervals. It is found that the coastline reaches the tip of the breakwater after 3.6 years, which agrees well with the estimate. The dashed lines show what happens later. There is a sand transport across the tip of the breakwater then, but it is less than the equilibrium transport, so the coastlines keeps changing, though at a lesser rate.

9.4 Exercises

1. Determine the diffusion coefficient if the following data are available:

   $a = 18$ m

   $\phi_0 = 0.2$ rad

   $s_0 = 1.5 \ 10^6$ m^3/year

2. Suppose a beach nourishment is carried out, i.e. a quantity of sand is deposited on the beach over a length L. How would you handle this in terms of initial or boundary conditions? Sketch the subsequent coastline development.

3. Determine the time step for the example of section 9.3 if you use the Crank–Nicholson method with $\theta = 0.5$.

Chapter 10
Consolidation of Soil

10.1 Mathematical Formulation

If soil, saturated with water, is loaded, e.g. by constructing a building on top of it, a deformation can occur, even if water and grains are considered incompressible. The reason is that the grains can move relative to one another, such that the pore volume changes. The water in the pores has to flow in or out, which takes some time. This process of consolidation is therefore time dependent. For a more comprehensive discussion see Verruijt (1983). Here, we consider a simple case where the process occurs in the vertical dimension only (Fig. 10.1).

If the change in volume of a soil element during a time interval Δt is ΔV and the vertical rate of flow per unit area is q, the water mass balance for the element is

$$\Delta V = -\frac{\partial q}{\partial z} \Delta z \, \Delta x \, \Delta y \, \Delta t = -\frac{\partial q}{\partial z} V \Delta t$$

A relative change of volume can be defined:

$$\Delta e = \frac{\Delta V}{V} = -\frac{\partial q}{\partial z} \Delta t$$

and by taking the limit you get the equation of continuity for the pore water

$$\frac{\partial e}{\partial t} + \frac{\partial q}{\partial z} = 0 \qquad (10.1)$$

The flow of groundwater depends on the local pressure gradient by the Darcy equation

$$q = -\frac{k}{\rho g} \frac{\partial p}{\partial z} \qquad (10.2)$$

in which

p = pressure in the groundwater
ρ = water density
k = permeability coefficient (compare eq. 7.2)

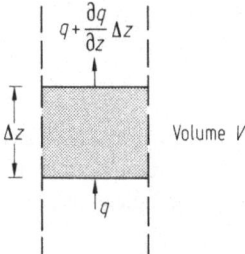

Fig. 10.1 Vertical flow of groundwater.

This comprises the equations of motion of the pore water. For the grains, you know that the volume does not change. If you consider the total normal stress σ in a horizontal section (e.g. the upper boundary of the element in Fig. 10.1), it is composed of two contributions:

$$\sigma = \sigma' + p \tag{10.3}$$

in which σ' is the stress between the grains. Something must be assumed for the mechanical properties of the system of grains. This is a quite complicated subject; for simplicity you might assume that it behaves like a simple elastic material (Hooke's law):

$$e = -m_v \sigma' \tag{10.4}$$

in which m_v is an elasticity coefficient, or rather (due to the minus sign) a coefficient of compressibility. This is not in contradiction with the incompressibility of the individual grains: eq. 10.4 concerns the total system of grains, which does respond to a force or stress.

Finally, the total stress σ should be determined. In this simple example, you can state that the column of soil as a whole is in equilibrium, so σ is a constant, given by the total load at the top. In fact, all stresses discussed here are stress increments relative to an equilibrium situation in which the stress balances the weight of soil and water.

You have now four equations for the four unknowns e, q, p, σ'. It is a simple exercise to derive one equation with one unknown, e.g.

$$\frac{\partial p}{\partial t} - D \frac{\partial^2 p}{\partial z^2} = 0 \tag{10.5}$$

This is a standard diffusion equation with diffusion coefficient

$$D = c_v = k/\rho g m_v \tag{10.6}$$

which is usually called the consolidation coefficient in soil mechanics. If you solve p from eq. (10.5), you can determine the relative volume change from eqs. (10.3) and (10.4), and finally the total absolute volume change which is usually the quantity you are interested in (as it determines the settling rate of your building).

10.2 Numerical Example

On a layer of 10 m of clay, saturated with water and in equilibrium, a layer of 2 m of dry sand is deposited. Under the layer of clay, the soil can be considered to be practically impermeable to water. The consolidation coefficient is $c_v = 10^{-6}$ m²/s. How will the surface level of the sand drop?

First of all you need initial and boundary conditions according to the rules given in section 7.1. In the initial situation, directly after the sand has been put on, the total stress (excess) equals the weight of the sand:

$$\sigma_0 = (1 - n)\rho_s \, gd = 3.5 \; 10^4 \, \text{N/m}^2$$

with a porosity $n = 0.35$, $d = 2$ m and density of sand $\rho_s = 2700$ kg/m³. The grains are still in their original position, so the change in interfacial stress between the grains is $\sigma' = 0$. The excess stress must therefore be provided by the water pressure:

$$p = \sigma_0 \text{ at } t = 0$$

At the bottom boundary ($z = -10$ m), the flow of water is zero, so due to eq. (10.2) $\partial p/\partial z = 0$. The upper boundary is taken at the top of the clay layer; it is assumed that water can leave the clay layer freely, so that the pressure remains constant ($p = 0$).

Next, you must specify the time step and grid interval, using the theory of chapter 8. At $t = 0$, the pressure takes a block form with length 10 m; due to the symmetry condition at the bottom, this corresponds to wave lengths of 20 m and larger. The total time of the simulation is not prescribed here; a typical time is the relaxation time for the wave length of 20 m:

$$t = (k^2 D)^{-1} = 10^7 \, \text{s} = 100 \text{ days}$$

As a condition of accuracy, you might require that numerical damping at this time is less than 1%. Using an implicit method with $\theta = 0.5$, you can figure out that this can be obtained with $k^2 D\Delta t \approx 1$; however, this results in an unrealistically large time step. Take, e.g., $\Delta t = 10$ days. A grid size $\Delta z = 1$ m is small enough.

The numerical treatment of the gradient boundary condition at the bottom can be done as follows. Define a virtual point $j = -1$ at $z = -10 - \Delta z$. The boundary condition leads to

$$p_{-1} = p_1$$

Applying this together with the finite-difference equation (7.12) at $j = 0$ gives a sufficient number of equations.

Some results are shown in Fig. 10.2 at 50-day intervals. You can see that the pressure change has indeed extended over the whole region after 100 days, but the process has not finished at that time. The total change of volume after 200 days is

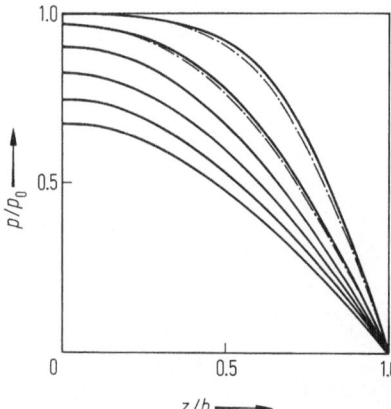

$z/h \longrightarrow$

Fig. 10.2 Computed pressure at 50-day intervals.

computed as

$$I = \int_0^h e\,dz = -m_v\left(\sigma_0 h - \int_0^h p\,dz\right)$$

or $\quad I/m_v\sigma_0 = h - \int_0^h p/\sigma_0\,dz = 4.647$

To check the numerical accuracy experimentally (which is a very useful thing to do in addition to the theoretical estimate), the same case has been treated much more accurately with $\Delta z = 0.5$ m and $\Delta t = 1$ day. The result is given by dashed lines in the figure. The difference is small so that the theoretical estimate of accuracy is confirmed. The total dimensionless change of volume is now 4.687.

Chapter 11
Convection–Diffusion

11.1 Transport of a Dissolved Substance

In Chapter 4, the transport of a dissolved substance was discussed without taking diffusion into account. This is not very realistic as there are a number of causes by which a substance (such as salt or a waste material but also temperature) is spread out in addition to being transported with the mean flow: molecular diffusion, turbulent mixing, and (very importantly) variations of flow velocity over the river cross-section. For a detailed discussion see, e.g., Fischer et al. (1979). You will find it intuitively plausible that a substance tends to be transported by these processes from locations where the concentration is high to locations where it is small, such as shown in Fig. 11.1. This is called diffusion here, whatever the cause. For the spreading by velocity variations, the term dispersion is often used, which has, however, other meanings as well. The transport with the mean velocity is called convection and the combined phenomenon convection–diffusion.

The transport of the dissolved substance across some cross-section correspondingly consists of two parts:

$$s = A\left(uc - D\,\frac{\partial c}{\partial x} \right) \tag{11.1}$$

where

A = area of cross-section

u = mean velocity

c = mean concentration in the cross-section

The diffusion coefficient D may vary strongly from one case to another. For a natural river, it has been found that

$$D = kac_f^{1/2}|u| \tag{11.2}$$

where a is the water depth and k an empirical factor that may vary from 50 to 500.

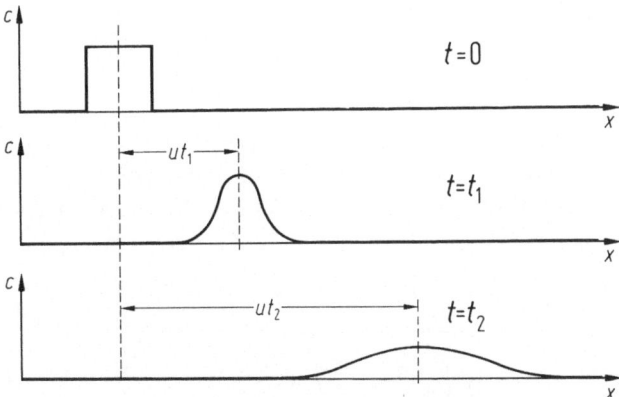

Fig. 11.1 Convection and diffusion.

The mass balance for the dissolved substance is

$$\frac{\partial}{\partial t}(Ac) + \frac{\partial s}{\partial x} = 0 \tag{11.3}$$

Substitute into this eq. (11.1):

$$\frac{\partial}{\partial t}(Ac) + \frac{\partial}{\partial x}(Qc) - \frac{\partial}{\partial x}\left(AD\frac{\partial c}{\partial x}\right) = 0 \tag{11.4}$$

Performing the differentiation, taking into account the water balance eq. (4.2) and assuming A and D to be constant, you get the "prototype" convection–diffusion equation

$$\frac{\partial c}{\partial t} + u\frac{\partial c}{\partial x} - D\frac{\partial^2 c}{\partial x^2} = 0 \tag{11.5}$$

This can be considered a combination of the simple-wave equation (4.3) and the diffusion equation (7.4) but mathematically it has the properties of the latter. For example, if you introduce a variable $z = x - ut$, you get the diffusion equation (please check)

$$\frac{\partial c}{\partial t} - D\frac{\partial^2 c}{\partial z^2} = 0 \tag{11.6}$$

Consequently, convection–diffusion effectively means diffusion in a frame of reference moving with the mean flow. You can expect that the same number and type of initial and boundary conditions is needed, i.e.

– one initial condition in each point of the region
– one boundary condition at each boundary, as a function of time.

11.2 Numerical Method

The numerical treatment of the convection–diffusion equation is very much like that of the pure diffusion equation. Using the same grid points in a six-point "molecule", you can write down the equivalent of the Crank–Nicholson method (section 7.3):

$$\frac{c_j^{n+1} - c_j^n}{\Delta t} + \theta \left\{ u \frac{c_{j+1}^{n+1} - c_{j-1}^{n+1}}{2\Delta x} - D \frac{c_{j+1}^{n+1} - 2c_j^{n+1} + c_{j-1}^{n+1}}{\Delta x^2} \right\} +$$

$$+ (1 - \theta) \left\{ u \frac{c_{j+1}^n - c_{j-1}^n}{2\Delta x} - D \frac{c_{j+1}^n - 2c_j^n + c_{j-1}^n}{\Delta x^2} \right\} = 0 \quad (11.7)$$

If $\theta = 0$, this is an explicit method of the FTCS type; otherwise, it is implicit. In the latter case, the system of difference equations, together with the boundary conditions, can be solved using the Thomas algorithm as in section 7.4.

In an exercise, the amplification factor of the method is derived:

$$\rho = \frac{1 + (1 - \theta)\lambda(\cos\xi - 1) - (1 - \theta)i\sigma\sin\xi}{1 - \theta\lambda(\cos\xi - 1) + \theta i\sigma\sin\xi} \quad (11.8)$$

with the, now well-known, parameters

diffusion parameter $\lambda = 2D\Delta t/\Delta x^2$

Courant number $\sigma = u\Delta t/\Delta x$

In the exercises, it is also shown that the method is unconditionally stable if $\frac{1}{2} \leqslant \theta \leqslant 1$ as before. Moreover, the explicit method · is stable under the two conditions

$$\sigma^2 \leqslant \lambda \leqslant 1 \quad (11.9)$$

In the explicit case, it is interesting to see which of the two conditions in eq. (11.9) is the most restricting. Given a grid size Δx, the right-hand inequality gives

$$\Delta t \leqslant \Delta t_1 = \Delta x^2/2D$$

The left-hand inequality turns out to be independent of the grid interval:

$$\Delta t \leqslant \Delta t_2 = 2D/u^2$$

The former is more restricting to the time step if

$$\Delta t_1 \leqslant \Delta t_2$$

or $|R| = |u\Delta x/D| \leqslant 2$

where you meet the *cell Reynolds number* R which tells something on the coarseness of the grid. Apparently, for a fine grid the usual diffusion criterion $\lambda \leqslant 1$ is the most important one. Later you will find that the cell Reynolds number also plays a part in the accuracy.

11.3 Application

In practical cases, the coefficients usually are not constant as supposed in eq. (11.5). Then, it is advisable to discretize the two basic equations (11.1) and (11.2) directly. Using the grid of Fig. 11.2, this can be done as follows:

The transports s are defined halfway between the grid points. Then the mass balance equation can be discretized as

$$\frac{(Ac)_j^{n+1} - (Ac)_j^n}{\Delta t} + \theta \frac{s_{j+1/2}^{n+1} - s_{j-1/2}^{n+1}}{\Delta x} + (1-\theta)\frac{s_{j+1/2}^n - s_{j-1/2}^n}{\Delta x} = 0 \tag{11.10}$$

The transports can be computed effectively using central differences:

$$s_{j+1/2}^n = A_{j+1/2}^n \left\{ \tfrac{1}{2} u_{j+1/2}^n (c_{j+1}^n + c_j^n) - D\frac{c_{j+1}^n - c_j^n}{\Delta x} \right\} \tag{11.11}$$

For the case of constant coefficients, this leads to the Crank–Nicholson method as given above.

Consider now the release of a quantity of salt in a channel, in which there is a constant mean flow of 0.02 m/s. The diffusion coefficient is estimated as $D = 1.6$ m^2/s. At a certain time you notice that there is a "cloud" of salt over a length of 1500 m. What will it look like when it arrives at 10 km downstream? This is simulated using a numerical model based on the Crank–Nicholson method with

$\theta = 0.55$

$\Delta x = 250$ m

$\Delta t = 5000$ s

Consequently, $\sigma = 0.4$ and $\lambda = 0.256$, so an explicit method would have worked as well. The result is shown in Fig. 11.3, indicating that the maximal concentration has decreased to about half its original value. If you had neglected the diffusion, the peak value would have remained constant.

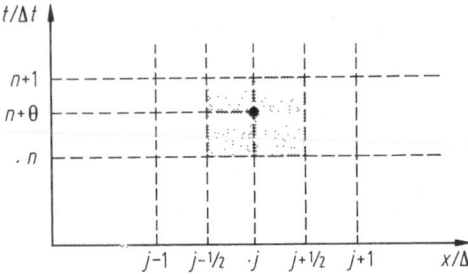

Fig. 11.2 Grid for balance equation.

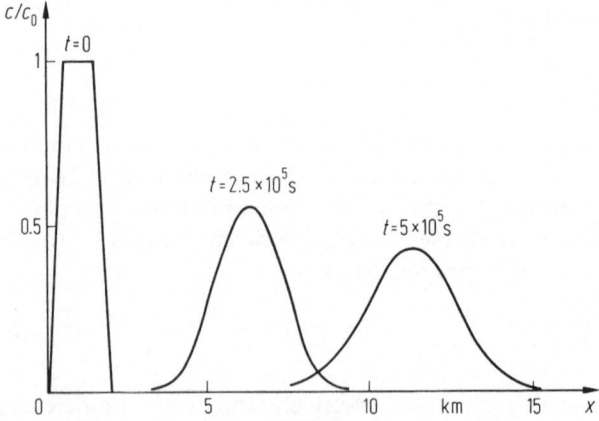

Fig. 11.3 Numerical simulation of salt transport.

11.4 Exercises

1. Formulate the boundary conditions for the convection–diffusion equation:

   – for BOD in a river flowing out of a lake in which a certain concentration of BOD is known;
   – for a closed end of a channel (or harbour basin)

2. Show that the amplification factor for the Crank–Nicholson method is given by eq. (11.8). Show that the method is unconditionally stable for $\frac{1}{2} \leqslant \theta \leqslant 1$.

3. Show that the FTCS method for the convection–diffusion equation is stable if $\sigma^2 \leqslant \lambda \leqslant 1$.

Chapter 12
Numerical Accuracy for Convection–Diffusion

12.1 Wave Propagation

In order to study the numerical accuracy for the convection–diffusion equation, you could follow the same approach as in Chapter 8. Unfortunately, this gets quite complicated, as an additional parameter (the Courant number) comes in. For practical purposes, it is often sufficient to consider diffusion and convection separately, and in that order. The procedure is then:

(a) determine relevant length and times scales for diffusion, as described in sections 8.1 and 8.2.
(b) determine the numerical parameters for diffusion as described in section 8.3.
(c) for the shortest relevant wave length found in step (a), determine the numerical parameters for pure convection, as described in section 5.5 (taking into account, of course, the correct speed of propagation). The idea is that shorter waves will be damped by diffusion, so there is no point in trying to "convect" them accurately. Moreover, the shortest wave is the critical one for numerical accuracy.
(d) use the smaller of the time steps and grid intervals, found in steps (b) and (c).

12.2 Example

In the example of section 11.3, the initial situation is a "block" of length 1500 m, so relevant wave lengths are 1500 m and larger. The transfer function at time 140 h (the travel time over 10 km) is such that the shorter waves are completely damped. The wave length still relevant at that time follows from

$$\exp\left(-Dk^2 t\right) \simeq \exp\left(-1\right)$$

which gives $L \simeq 5600$ m.

Using this wave length, the time step for the diffusion process follows from eq. (8.10). With some experimentation, you find that $k^2 D \Delta t = 0.1$ is sufficiently small. This means that time steps up to 14 h might be used. For the grid size, the relevant wave length is again 5600 m. A grid size of 250 m will be sufficiently small. In the initial situation, you still have 6 grid points per wave length, which is not very large but acceptable.

Finally, convection with speed 0.02 m/s is considered. The "wave period" for the wave considered is $T = L/u = 2.8 \times 10^5$ s. You find for the wave damping and relative speed of propagation:

$T/\Delta t$	n	d_n $\theta = 0.55$	d_n $\theta = 1$	c_r $\theta = 0.55$	c_r $\theta = 1$
10	18	0.72	0.05	0.97	0.89
20	36	0.84	0.18	0.992	0.97
50	90	0.93	0.49	0.999	0.995
100	180	0.97	0.70	0.9997	0.999

In the wave speed, there is not much of a problem. The errors in the wave amplitude are much more severe than for diffusion. Obviously, $\theta = 1$ is too inaccurate. For $\theta = 0.55$, reasonable values are $T/\Delta t = 50$ or 100, which gives time steps of about 5000 to 10 000 s. These values have been used in numerical simulations shown in Fig. 12.1. The conclusions on wave damping and phase shift are confirmed by the numerical results. The numerical parameters used in section 11.3 are found to be acceptable.

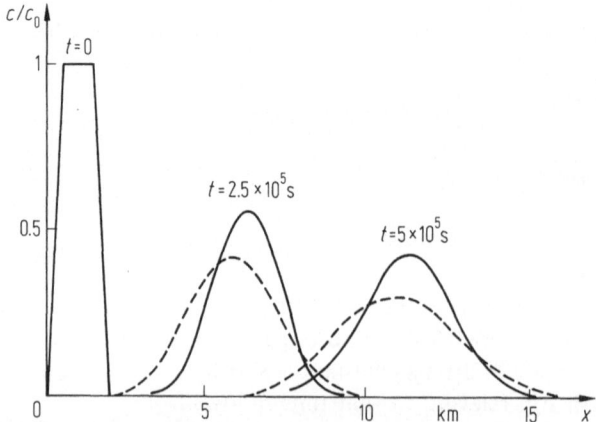

Fig. 12.1 Salt transport in a channel. Drawn lines: $\Delta t = 5000$ s, $\theta = 0.55$; dashed lines: $\Delta t = 10000$ s, $\theta = 1$.

12.3 Numerical Diffusion

The analysis of the truncation error is of particular importance in the case of convection-diffusion. As an example, consider the FTCS method (eq. 11.7 with $\theta = 0$). Developing this into Taylor series relative to the point (j, n), you find (please check):

$$\frac{\partial c}{\partial t} + u\frac{\partial c}{\partial x} - D\frac{\partial^2 c}{\partial x^2} = -\frac{1}{2}\Delta t\frac{\partial^2 c}{\partial t^2} - \frac{1}{6}u\Delta x^2\frac{\partial^3 c}{\partial x^3} + \frac{1}{12}D\Delta x^2\frac{\partial^4 c}{\partial x^4} \tag{12.1}$$

This is first order in time and second order in space. The time derivative can be converted into spatial derivatives as follows:

$$\frac{\partial^2 c}{\partial t^2} = -\frac{\partial}{\partial t}\left(u\frac{\partial c}{\partial x} - D\frac{\partial^2 c}{\partial x^2}\right) = u^2\frac{\partial^2 c}{\partial x^2} - 2uD\frac{\partial^3 c}{\partial x^3} + D^2\frac{\partial^4 c}{\partial x^4} \tag{12.2}$$

(supposing u and D to be constants). The first term on the right-hand side is of the same form as the diffusion term; it is therefore called numerical diffusion with coefficient

$$D_{num} = -\tfrac{1}{2}u^2\Delta t \tag{12.3}$$

This does not mean that the other terms of eq. (12.2) are unimportant (they are of the same order) but they are interpreted less simply. In this case you notice that the numerical diffusion coefficient is negative. This indicates a tendency towards instability as you can conclude from eq. (8.5). To prevent this, the total diffusion coefficient should at least be positive:

$$D + D_{num} > 0 \tag{12.4}$$

which results in

$$\Delta t < 2D/u^2 \tag{12.5}$$

This condition has been found before in eq. (11.9). However, this is of course not sufficient. From an accuracy point of view you want the apparent diffusion coefficient $D + D_{num}$ to be near the physical one D; otherwise you are solving the wrong problem. This means that Δt should be *much* smaller than the value shown in eq. (12.5), which usually leads to rejection of the FTCS method.

Please note carefully that this numerical diffusion originates from a time derivative, so it works only in a nonsteady situation. This may sound strange (cf. Roache, 1976) as you might think that the numerical diffusion term will work in a steady state as well. Actually, it will not be zero, but the three terms on the right-hand side of eq. (12.2) together are zero.

In an exercise, it is found that the numerical diffusion coefficient for the Crank–Nicholson method is

$$D_{num} = (\theta - \tfrac{1}{2})u^2\Delta t \tag{12.6}$$

This is at least positive if $\theta > \tfrac{1}{2}$, in agreement with the stability analysis. Its magnitude for $\theta = 1$ is, however, equal to that of the FTCS method, apart from the sign. Therefore, it is usually not a very good idea to use that value of θ. If you take $\theta = \frac{1}{2}$, which is just acceptable from a stability point of view, the numerical diffusion term is zero. However, do not make the mistake of thinking that there is no numerical error: the other terms of eqs. (12.1) and (12.2) are nonzero.

12.4 Example

For the two cases of Fig. 12.1, the numerical diffusion coefficients can be computed as 0.1 and 2 m^2/s, respectively. As the physical diffusion coefficient is 1.6 m^2/s, the latter case is clearly unacceptable. The effect of the numerical diffusion can be clearly seen in Fig. 12.1.

12.5 Convection Only

The phenomenon of numerical diffusion occurs for $D = 0$ as well. As an example, consider the modified Lax method for the simple-wave equation (5.10). Applying eq. (12.2) again, you find

$$\frac{\partial c}{\partial t} + u \frac{\partial c}{\partial x} = \frac{\Delta x^2}{2\Delta t} (\alpha - \sigma^2) \frac{\partial^2 c}{\partial x^2} + \dots \tag{12.7}$$

with a numerical diffusion coefficient

$$D_{num} = (\alpha - \sigma^2) \, \Delta x^2 / 2\Delta t$$

There is no physical diffusion to balance a possible negative numerical one, so for stability it is required that

$$\sigma^2 < \alpha$$

in (partial) agreement with eq.(5.5). Note that in this way you find some but not all of the stability conditions.

   To show that the numerical diffusion coefficient indeed works that way, the modified equation (12.7) has been solved analytically with the following data (Vreugdenhil 1969):

   initial condition $c = 0$

   boundary condition at $x = 0$: $c = c_0$
   $u = 1$

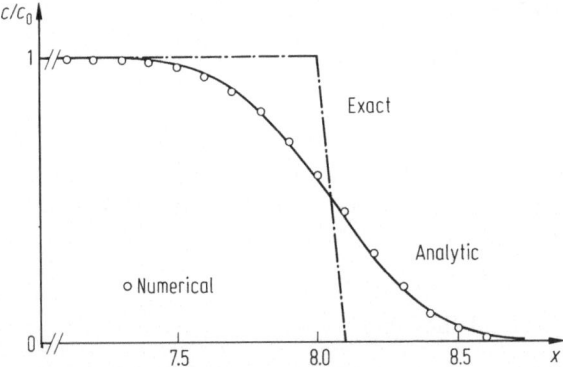

Fig. 12.2 Numerical diffusion in the simple-wave equation, together with analytic interpretation. (with permission from Vreugdenhil, 1969)

$$\alpha = 0.9$$
$$\Delta x = 1$$
$$\Delta t = 0.9$$

which gives $D_{num} = 5$ (the values are more or less arbitrary for demonstration purposes). Figure 12.2 shows the analytic solution at $t = 8$ together with the numerical result using the modified Lax method.

12.6 Wiggles

In some cases, numerical solutions of the convection-diffusion equation show numerical "wiggles" or short waves with wave length $2\Delta x$. They are clearly of numerical origin. Due to a regrettable misunderstanding in Roache (1976), they have for a long time been attributed to instability; however, they have nothing to do with that (Hindmarsh et al., 1984). Actually, the wiggles do not grow in time and even persist in a steady state (Fig. 12.3). A very useful discussion of wiggles has been given by Gresho and Lee, (1979).

The occurrence of wiggles can be demonstrated quite easily for the steady-state convection-diffusion equation:

$$u \frac{\partial c}{\partial x} - D \frac{\partial^2 c}{\partial x^2} = 0 \tag{12.8}$$

Both the FTCS and Crank–Nicholson methods (and, as a matter of fact, a number of other numerical methods) in this case reduce to

$$u\frac{c_{j+1}-c_{j-1}}{2\Delta x} - D\frac{c_{j+1}-2c_j+c_{j-1}}{\Delta x^2} = 0 \tag{12.9}$$

The time discretization has nothing to do with it. Suppose you want to solve this equation for a river carrying a concentration c_0 of some substance and flowing into a lake. In the lake water, the substance does not occur. The boundary conditions for eq. (12.8) then read:

$c = c_0$ at $x = 0$ (far upstream)

$c = 0$ at $x = L$ (at the lake)

The analytic solution is composed of exponential functions $\exp(\mu x)$. Substitution shows that the two possible values for μ are 0 and u/D. Taking the boundary conditions into account, you find the solution:

$$c = c_0\left\{1 - \frac{1-\exp(ux/D)}{1-\exp(uL/D)}\right\} \tag{12.10}$$

The exact solution of the finite-difference equation (12.9) is found in a similar way; you have to be aware that in the case of finite differences the role of exponential functions is taken by

$$c_j = r^j$$

Substituting this into eq. (12.9) you find this to be possible if

$$(\tfrac{1}{2}R - 1)r^2 + 2r - (\tfrac{1}{2}R + 1) = 0$$

where $R = u\Delta x/D$ is called the cell Reynolds number. It is also indicated as the cell Péclet number, which is actually more correct in this case. There turn out to be two possible values (corresponding to the two exponential functions in the analytic case):

$$r_1 = 1 \qquad r_2 = (2+R)/(2-R) \tag{12.11}$$

One part is a constant, as it should; the other part behaves like an exponential function. A combination of the two solutions satisfying the boundary conditions is:

$$c_j = c_0\{1 - (1-r_2^j)/(1-r_2^J)\} \tag{12.12}$$

where J is the total number of grid points. You can easily check that this is a good approximation of eq. (12.10) if R is small (even as large as 1). However, if $R > 2$, a qualitative difference occurs, as r_2 gets negative and the numerical solution oscillates. Both cases are shown in Fig. 12.3.

From this discussion, it is obvious that the numerical oscillations in this case are not related to stability but to accuracy. The condition $R < 2$, required to avoid the oscillations is just a condition for the grid size to be sufficiently small.

To complicate things, the wiggles do not always show up, even if $R > 2$. This is shown in Fig. 12.3 where the downstream boundary condition has been changed into $\partial c/\partial x = 0$, which ignores the influence of the lake concentration on the river (this may be a quite good approximation if the concentration does not intrude into

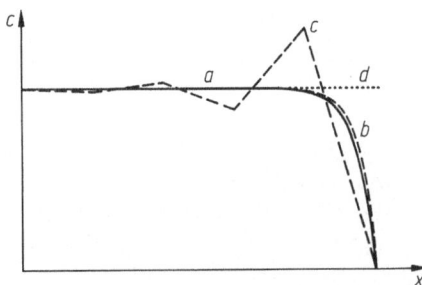

Fig. 12.3 Numerical Oscillations for uL/D $= 20$; *a*. analytic solution. *b*. numerical solution for $R = 1$. *c*. numerical solution for $R = 4$. *d*. the same with downstream boundary condition $\partial c/\partial x = 0$.

the river very far). In this case, the wiggles could exist, but they are not excited by the boundary conditions. In general, you can expect this sort of wiggles in areas with strong concentration gradients.

Some people have attempted to circumvent the wiggles by using upstream (or upwind) methods. The convection term is then approximated by a backward difference (opposite the flow direction), i.e. if $u > 0$:

$$u\frac{c_j - c_{j-1}}{\Delta x} - D\frac{c_{j+1} - 2c_j + c_{j-1}}{\Delta x^2} = 0 \qquad (12.13)$$

The numerical diffusion coefficient is determined in an exercise:

$$D_{num} = \tfrac{1}{2}|u|\Delta x \qquad (12.14)$$

The accuracy is only first order and you get a very inaccurate solution. It is, however, free from wiggles (Fig. 12.4). The method can be used, but only with utmost care due to the low accuracy. Unfortunately, several applications of this method have been reported in the literature that are far from the truth (as in Fig. 12.4). Higher-order upstream methods exist, that are more accurate than the first-order one shown here; they are less sensitive to, but not free from numerical oscillations (e.g. Leonard, 1980). Moreover, they involve one more grid point on the upstream side, which makes things more complicated, particularly near the boundaries.

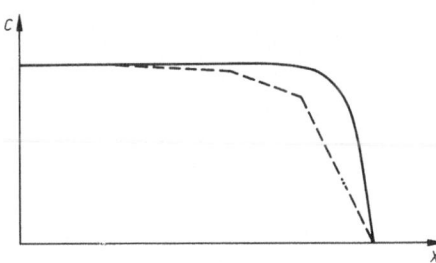

Fig. 12.4 Numerical solution with upstream method for $R = 4$.

12.7 Exercises

1. Derive the numerical diffusion coefficient (12.6) for the Crank–Nicholson method. Hint: take the reference point as $(j\Delta x, (n+\theta)\Delta t)$. Note that the only term contributing to numerical diffusion is the time derivative.
2. Derive the numerical diffusion coefficient (12.14) for the explicit upstream method. You will draw a remarkable conclusion by computing the effective cell Reynolds number taking the total (physical + numerical) diffusion coefficients into account and showing that it never exceeds 2. This explains why you do not get wiggles. However, what is the ratio between the total and physical diffusion coefficients and what does this mean for accuracy?
3. Find the exact solution to the upstream method, to be compared with eq. (12.10). Show that no oscillations can occur, whatever the value of the cell Reynolds number (investigate the sign of r_2).
4. Write the upstream finite-difference equation for the case that $u < 0$.
5. In general, a three-point finite-difference method for the steady-state convection-diffusion equation can be written as

$$\alpha_{j-1} c_{j-1} + \alpha_j c_j + \alpha_{j+1} c_{j+1} = 0$$

Show that for consistency with the differential equation the sum of the coefficients α_j should be zero (Taylor expansion). Show that, in order to prevent numerical oscillations, α_{j-1} and α_{j+1} should have the same sign. Check whether the central and upstream difference methods satisfy this.

Chapter 13
Salt Intrusion in Estuaries

13.1 Formulation

In a tidal river, salt water from the sea tends to penetrate due to its slightly greater density. The form in which this happens depends on the tidal influence. If there is hardly any tide, the salt water will penetrate underneath the fresh river water as a "salt wedge". Far more common is the situation where a strong tidal action takes care of a mixing process, such that the water in a particular cross-section has an almost uniform salt concentration; however, this concentration varies in long-itudinal direction. This is called the well-mixed case. The salt concentration is now governed by the convection-diffusion equation (11.4) with two complications:

(i) the hydrodynamic variables A and Q are now functions of space and time, to be computed as discussed in Chapter 16;
(ii) the diffusion coefficient is influenced by the longitudinal gradient of salt concentration. This is a rather complicated matter (cf. Thatcher and Harleman 1972), that is not discussed here.

Once the situation is in dynamic equilibrium (i.e. more or less repeats itself in each tidal period), there will be a mean concentration gradient such that the diffusive transport balances the convective transport by the mean river flow u_0 and (almost) steady. Taking the diffusion coefficient constant as an approximation, this means

$$u_0 \frac{\partial c}{\partial x} - D \frac{\partial^2 c}{\partial x^2} = 0 \qquad (13.1)$$

The concentration pattern generated by this equation is carried back and forth by the tidal flow in a periodic manner, over a distance called the tidal excursion. This rough picture of the process can be used for the investigation of accuracy in the next section.

The initial condition for the salt concentration is unknown. You may choose an estimated distribution (or zero) and run the model for a number of tidal cycles until the situation repeats itself. This may take quite some time, so a reasonable guess is of some importance.

The upstream boundary condition is simple: zero salt concentration. The downstream one, again, results in a complication. During inflow, there is no problem, as you can prescribe the concentration of the inflowing sea water. However, during outflow this is no longer correct. Actually, from a physical point of view you would not want to apply any boundary condition at all; the concentration at the boundary is determined by the water flowing out. Yet, a boundary condition is needed. You will look for a "harmless" condition, that does not fix the situation but satisfies the mathematical need for a boundary condition. A useful condition is

$$\frac{\partial^2 c}{\partial x^2} = 0 \qquad (13.2)$$

Note that putting the first derivative equal to zero is not so good, as this would mean that the diffusive transport in eq. (11.1) is zero. With condition (13.2), the concentration drops during outflow. At the turn of the tide (from outflow to inflow again) you would get a jump in the concentration from the low value of outflowing water to the high sea-water concentration. Therefore, this transition is supposed to take some time, during which the concentration rises linearly (Fig. 13.1). The time needed for the transition has to be determined from field measurements.

As an example, take a river with the following data:

cross-section $\qquad$ $A_s = 3000 \text{ m}^2$

river flow $\qquad$ $u_0 = 0.07 \text{ m/s}$

tidal velocity $\qquad$ $u_1 = 0.7 \text{ m/s (amplitude)}$

diffusion coefficient $\qquad$ $D = 300 \text{ m}^2/\text{s}$

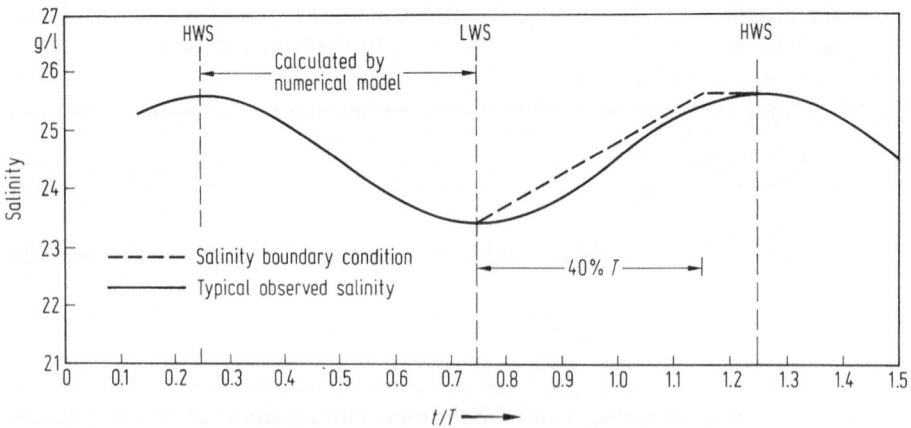

Fig. 13.1 Concentration of salt at seaward boundary.

13.2 Accuracy Mean Concentration

For an analysis of the mean concentration, eq. (13.1) can be integrated once. As there is no net salt flux, you get

$$u_0 c - D \frac{\partial c}{\partial x} = 0 \qquad (13.3)$$

The analytic solution of this is

$$c(x) = c_0 \exp(u_0 x/D) \qquad (13.4)$$

(note that $u_0 < 0$ if the positive x-direction is upstream). With the values given, the intrusion (relaxation) length is $D/u_0 = 4.3$ km.

The numerical representation of this, with the methods discussed in Chapter 11, is

$$\frac{1}{2} u_0 (c_j + c_{j+1}) - D \frac{c_{j+1} - c_j}{\Delta x} = 0 \qquad (13.5)$$

With an analysis similar to section 12.4, this gives

$$\tfrac{1}{2} R(r+1) - (r-1) = 0$$

or

$$r = (1 - R/2)/(1 + R/2)$$

Try some values of Δx and accept an error at a distance of one relaxation length of, say, 5%

Δx (km)	R	n	r^n	$\exp(u_0 x/D)(x = 4\,\text{km})$
4	0.93	1	0.364	0.393
2	0.47	2	0.384	
1	0.23	4	0.392	

This shows that a grid size of about 2 km is sufficient to represent the steady-state salt distribution.

13.3 Accuracy for Tidal Fluctuation

In the tidal fluctuation, it can be assumed that diffusion is not very important; the salt distribution is just convected with the tidal velocity. The travelling distance during half a tidal cycle is

$$x = \int_0^{T/2} u\,dt = u_1 \int_0^{T/2} \sin 2\pi t/T\,dt = u_1\,T/\pi$$

For simplicity, you could consider the equivalent situation where the salt is transported with a constant velocity u_1 during a time $T/\pi = 4\,\text{h}$.

The wave length can be estimated as twice the relaxation length for the steady-state distribution, i.e. about 9 km.

For the Crank–Nicholson method, applied to the simple-wave equation (which is what remains of the convection diffusion equation by neglecting diffusion), you can now try some values of Δx and Δt; assume $\theta = 0.55$. For a number of n time steps during 4 hours, you find

$\Delta x\,(\text{km})$	$\Delta t\,(\text{s})$	σ	ξ	n	d_n	c_r
4	1800	0.32	2.79	8	0.999	0.122
2	1800	0.63	1.40	8	0.869	0.683
1	1800	1.26	0.70	8	0.798	0.873
4	900	0.16	2.79	16	0.998	0.122
2	900	0.32	1.40	16	0.927	0.700
1	900	0.63	0.70	16	0.882	0.908

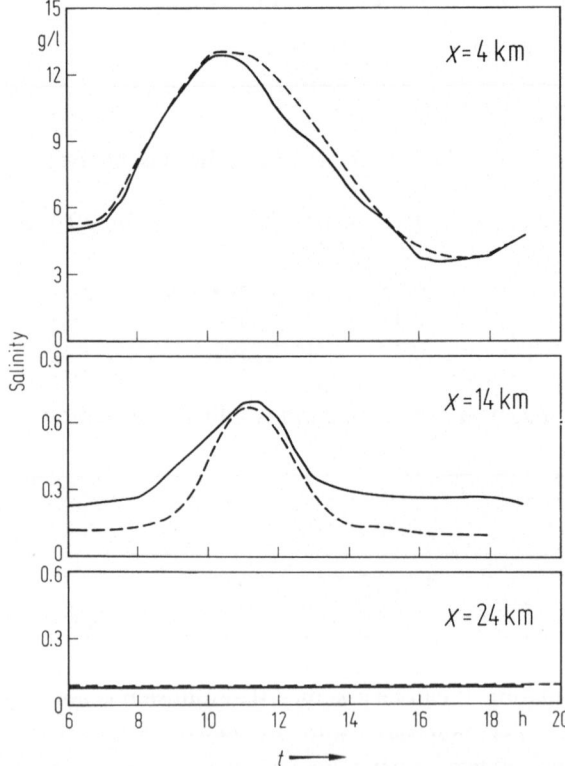

Fig. 13.2 Computed and measured salt concentrations at 4, 14 and 24 km from the sea. Drawn lines computed, dashed lines measured.

Reducing the grid size at constant time step improves the phase error, but at the same time gives a greater damping (this is because the Courant number increases). Reducing the time step at constant grid size results in a slight improvement of both amplitude and phase errors. For a modest accuracy, $\Delta x = 2$ km and $\Delta t = 900$ s is sufficient. Combining this with the result of the previous section, you see that $\Delta x = 2$ km can be used.

You should be well aware that all the approximations made in the latter two sections were intended only to simplify the analysis of numerical accuracy. The actual computations are made with the full equations, non-constant coefficients etc. You can expect that the numerical accuracy works out about the same way for this complete set of equations.

Figure 13.2 gives some results for a simulation with these values, for a real river in which the tidal flow was computed with a numerical model as well. Measured salt concentrations are included. The agreement is not ideal, but acceptable considering the semi-empirical diffusion coefficient used.

Chapter 14
Boundary Layers

14.1 Suspended Sediment Transport

The idea of boundary layers can be illustrated very well using the example of sediment transport in suspension. If water in a river flows over a sandy bottom at sufficiently high velocity, sand will be picked up and carried with the flow, even though it tends to fall back. The process by which sediment particles are kept in suspension is turbulent diffusion. Just as in chapter 11, turbulence causes an effective transport from regions with high sand concentration (near the bottom) to those with low concentration (near the water surface). If you consider a steady-state situation, the mass balance for suspended sand is

$$u\frac{\partial c}{\partial x} + \frac{\partial s}{\partial z} = 0 \tag{14.1}$$

in which the sediment flux in vertical direction is

$$s = w_s c - D\frac{\partial c}{\partial z} \tag{14.2}$$

where x and z are the horizontal and vertical coordinates, D is the turbulent diffusion coefficient and w_s the settling velocity or fall velocity of the sand particles. The latter is a property of the sand (particularly of the particle diameter) and it can be measured. Some assumptions have been made here:

– the flow is steady and uniform with a velocity distribution $u(z)$;
– the sediment flux due to turbulence is in vertical direction only. In reality, there is a similar flux in the longitudinal direction, but this will be much smaller as variations of concentration occur over much larger distances.

Combining the two equations, you get

$$u\frac{\partial c}{\partial x} + w_s\frac{\partial c}{\partial z} - \frac{\partial}{\partial z}\left(D\frac{\partial c}{\partial z}\right) = 0 \tag{14.3}$$

which is very much like the convection–diffusion equation (11.5). If you take the diffusion coefficient constant for a while and divide the equation by u, the formal equivalence with (11.5) is more easily seen:

$$\frac{\partial c}{\partial x} + \frac{w_s}{u}\frac{\partial c}{\partial z} - \frac{D}{u}\frac{\partial^2 c}{\partial z^2} = 0 \tag{14.4}$$

The longitudinal space coordinate x takes the place of time, w_s/u that of the convection velocity and D/u that of the diffusion coefficient. Therefore, you can use all methods discussed for the convection–diffusion equation, with the appropriate change of interpretation. Actually, the division by the flow velocity is not even necessary (see exercise No. 1 in section 14.6).

14.2 Example

Suppose that water with some arbitrary sediment concentration flows into a river with a bottom of fine sediment. Sand will be picked up or deposited and at a certain distance, an equilibrium situation is reached, where the vertical sand flux is zero. This means that

$$c(z) = c_0 \exp\left(-zw_s/D\right) \tag{14.5}$$

where c_0 is the concentration just above the bottom.

How large is this adaptation distance? You can estimate the order of magnitude as follows. Diffusion over a depth a with a diffusion coefficient D takes a time of the order a^2/D. This corresponds to a distance $a^2 u/D$. To get a better estimate, a numerical simulation can be used.

For the sediment transport equation (14.3), you need initial and boundary conditions. The "initial" condition is here $c = f(z)$ at $x = 0$. The boundary condition at the water surface is that the flux is zero: no sediment enters or leaves the region across the surface.

$$w_s c - D\frac{\partial c}{\partial z} = 0 \text{ at } z = a$$

The most difficult one is the boundary condition at the river bottom. The mechanism of sand pick-up is poorly understood and a good formulation of it in terms of sand and flow properties is not known. A common assumption is that the concentration right at the bottom adjusts itself very quickly to the flow conditions, so it is at its equilibrium value c_0 which is known experimentally. Then the bottom boundary condition is

$$c = c_0 \quad \text{at } z = 0$$

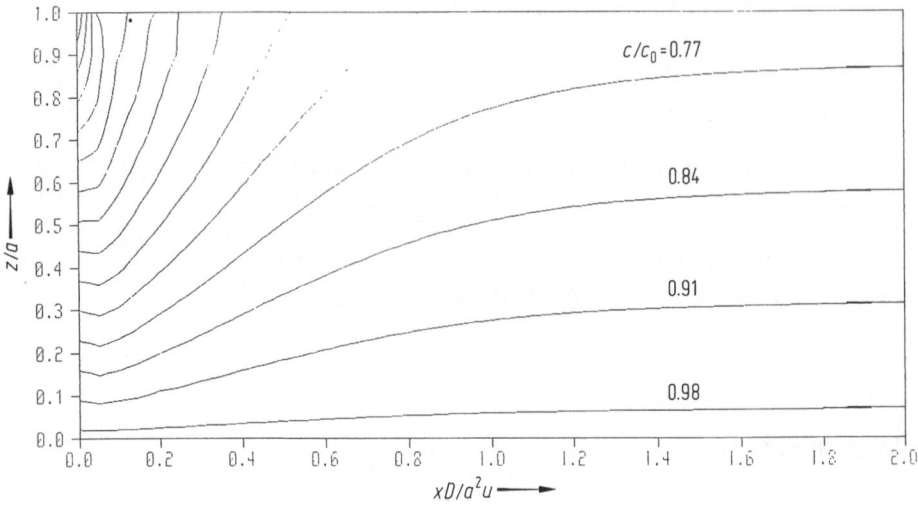

Fig. 14.1 Suspended sediment concentration (lines of constant c/c_0).

You should be aware that things are a little more complicated if the structure of turbulent flow is taken into account in more detail; here for simplicity it is assumed that the diffusion coefficient is constant and the velocity profile parabolic. This corresponds to "quasi-laminar" flow.

With these boundary conditions, the problem can be solved numerically. As the flow velocity gets quite small near the bottom, it is advantageous to use an implicit method (see exercise No. 2). The results shown in Fig. 14.1 have been obtained with the Crank–Nicholson method and the following data (note that the grid size Δx is expressed in terms of the estimated adaptation length):

$$u = u_0\, z(2a - z)/a^2$$
$$f(z) = c_0(1 - z/a)$$
$$w_s a/D = 0.3$$
$$\Delta z/a = 0.1$$
$$\Delta x D/u_0 a^2 = 0.05$$
$$\theta = 0.5$$

The equilibrium values according to eq. (14.5) are also indicated and it is seen that the numerical results converge to them. The adaptation length agrees very well with the estimate given above. You will note some oscillations in the upper left-hand corner. These originate from the fact that the initial condition does not satisfy the surface boundary condition. However, after a few steps the oscillations are damped out.

14.3 Boundary-Layer Flows

There are many situations like the one of the preceding section, where variations are much slower in the flow direction than across it. Examples are flow in a pipe, a river or along a fixed wall, a jet issuing from an outlet, or a plume from a chimney. Although only some of these have anything to do with a boundary, they are all boundary-layer type of flows. For an extensive treatment see, e.g., Cebeci and Smith (1974). In Fig. 14.2, the length scales over which some velocity variation occurs are indicated. In a boundary layer, $L/\delta \gg 1$.

With a little thought, you will see that the order of magnitude of the velocity components must behave the same way. Formally, this follows from the equation of continuity for an incompressible fluid:

$$\frac{\partial u}{\partial x} + \frac{\partial w}{\partial z} = 0 \tag{14.6}$$

If U, W indicate the order of magnitude of the velocity components (or actually their variations), the terms have the order of magnitude:

$$\frac{\partial u}{\partial x} \approx U/L \qquad \frac{\partial w}{\partial z} \approx W/\delta \tag{14.7}$$

so

$$U/W = L/\delta \tag{14.8}$$

This is the first step in deriving the boundary-layer approximation. The second step considers the momentum equation in the cross-flow direction (here the z-direction)

$$\frac{\partial w}{\partial t} + u\frac{\partial w}{\partial x} + w\frac{\partial w}{\partial z} = -\frac{1}{\rho}\frac{\partial p}{\partial z} - g + v\left(\frac{\partial^2 w}{\partial x^2} + \frac{\partial^2 w}{\partial z^2}\right) \tag{14.9}$$

where p is the pressure, g the acceleration due to gravity, and v the kinematic viscosity coefficient. The latter is treated as a constant for simplicity. If the flow is turbulent, this is not true, but the argument is approximately the same. Only steady flows are considered, so the first term drops out. As w is small, all terms involving it will be small compared with the gravity term. This can be demonstrated more formally by estimating the order of magnitude of each term, but for the present purpose you may accept the conclusion that gravity is balanced almost completely

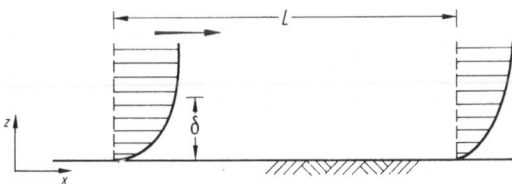

Fig. 14.2 Length scales in a boundary layer.

by the pressure gradient:

$$\frac{\partial p}{\partial z} = -\rho g \tag{14.10}$$

This indicates that in a boundary layer you will have a hydrostatic pressure distribution. A direct consequence is that $\partial p/\partial x$ does not depend on z (check this by differentiating eq. 14.10).

As a third and final step, consider the momentum equation in the flow direction:

$$\frac{\partial u}{\partial t} + u\frac{\partial u}{\partial x} + w\frac{\partial u}{\partial z} = -\frac{1}{\rho}\frac{\partial p}{\partial x} + v\left(\frac{\partial^2 u}{\partial x^2} + \frac{\partial^2 u}{\partial z^2}\right) \tag{14.11}$$

For steady flows, the first term again drops out. Apart from this, no important simplifications are possible in this equation, except in the viscosity terms. The latter of these will be much larger than the former because of the great difference in length scales: vertical gradients are much larger than horizontal ones (of the same variable!), and the more so for second derivatives. Therefore, the momentum equation in its boundary-layer form reads:

$$u\frac{\partial u}{\partial x} + w\frac{\partial u}{\partial z} - \frac{\partial}{\partial z}\left(v\frac{\partial u}{\partial z}\right) = -\frac{1}{\rho}\frac{\partial p}{\partial x} \tag{14.12}$$

To include the case of turbulent flow, the viscosity has been put inside the differential, so that it can be treated as a variable. Equation (14.12) is very much like eq. (14.3), although there the vertical flow velocity is exactly zero and replaced by the particle fall velocity. In eq. (14.12) you should realize that the first two terms have the same order of magnitude. The first has u^2 as opposed to uw in the second, but on the other hand the gradient in the z-direction is much larger than that in the x-direction and both effects roughly balance one another.

If you would know the pressure gradient, eq. (14.12), is a convection–diffusion equation just as eq. (14.3), now with u as unknown. Once you have solved it, the other component w can be solved from the equation of continuity (14.4). The pressure gradient (which is not a function of z, so just a parameter in the equation) has to be fixed by additional requirements.

14.4 Pressure Gradient

To determine the pressure gradient, you should take into account the boundary conditions. Firstly, consider a "real" boundary layer along a fixed wall. Suppose that the flow velocity far from the wall is known. Then the "far-away" boundary conditions for the boundary layer are

$$u = u_e, \qquad p = p_e \quad \text{at } z \to \infty \tag{14.13}$$

"Infinity" should not be taken literally: you can take two or three times the expected boundary-layer thickness.

At the fixed wall, there are two conditions as well:

$$w = 0, \qquad u = 0, \quad \text{at } z = 0 \tag{14.14}$$

The former is a kinematic condition, saying that no fluid passes through the wall. The latter condition is a dynamic one and it means that the fluid "sticks" to the wall due to viscosity; it is called the no-slip condition.

Collecting the boundary conditions, you will find that you have one in excess: two are needed for eq. (14.12) (being a diffusion equation), one for eq. (14.6). The additional condition is the one that fixes the pressure gradient. In this case, it is very simply equal to the (specified) pressure gradient outside the boundary layer. Consequently, the boundary-layer equation (14.12) has a known right-hand side and you can handle it like a normal convection–diffusion equation as in section 14.1.

The second case is a little more complicated. Consider the flow in a river or canal. At the river bottom, which is a fixed wall, you have the same conditions as before (eq. 14.14). For many applications, the water surface can be assumed known (e.g. being at its equilibrium position). Then at the surface the normal velocity component is zero (no fluid passes the surface, or: the flow is parallel to the surface):

$$u \frac{\partial h}{\partial x} - w = 0 \qquad \text{at } z = h \tag{14.15}$$

Secondly, as a dynamic condition, the shear stress is zero, as the free surface does not produce any friction (in absence of wind over the water surface). This gives

$$\frac{\partial u}{\partial z} = 0 \qquad \text{at } z = h \tag{14.16}$$

In fact, you have now replaced the free surface by what is called a "rigid lid", i.e. a frictionless lid on top of the water layer. The water is flowing through a hypothetical "tunnel" and doing so, it will produce a nonzero pressure on the rigid lid. In reality, this would cause a change of position of the free surface until you have a zero (or rather atmospheric) pressure, but the introduction of the lid means that you can no longer require that the pressure is zero. The boundary conditions are, therefore, (14.14) through (14.16). Again, there is one in excess, but unfortunately, none of them explicitly specifies the pressure.

If you integrate the continuity equation (14.6) over the water depth you get

$$\frac{\partial q}{\partial x} - \left[u \frac{\partial h}{\partial x} - w \right]_{z=h} = 0 \tag{14.17}$$

in which q is the discharge per unit width and the first condition of (14.14) has been used. Physically speaking, the discharge must be constant and you see that this is only so if eq. (14.15) is satisfied and conversely. This boundary condition may, therefore, be replaced by the condition of a constant discharge.

In a more or less constructive manner, you can imagine the situation as follows. If you know the flow up to a section x, you can determine the velocity distribution

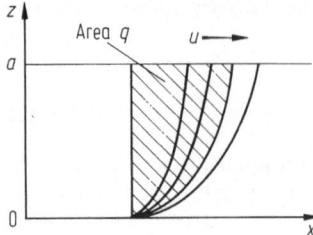

Fig. 14.3 Only one of the possible velocity profiles has the correct discharge.

at $x + \Delta x$ for a number of values of the pressure gradient (Fig. 14.3) using eq. (14.16) and the second of eq. (14.14). Then you may determine w using the first of eq. (14.14) as a boundary condition. Only one of these distributions will have the correct discharge i.e. satisfies eq. (14.15). The pressure gradient turns out to be fixed by a continuity constraint. Numerically, you have to find it iteratively in the very same way.

14.5 Developing Flow in a River

In a river with a certain bottom slope S, the equilibrium flow profile is found from eq. (14.12) with the gravity term added and all x-derivatives zero:

$$-\frac{\partial}{\partial z}\left(v\frac{\partial u}{\partial z}\right) = gS \qquad (14.18)$$

For constant viscosity (in case of turbulent flow you could take an average value) this gives (please check)

$$u(z) = -\frac{gS}{2v}z(z - 2a) \qquad (14.19)$$

This is the well-known parabolic velocity distribution of (quasi-)laminar flow. However, this profile exists only if the flow is fully developed. If the flow comes into the river with some odd velocity profile (due to the inflow conditions), what is the distance needed to get the equilibrium profile? This is the same question as in section 14.2, but now for the velocity distribution.

The numerical method of solution has already been indicated. For eq. (14.12), you could use the FTCS or Crank–Nicholson methods. As you will see from exercise 2, the FTCS method suffers from very strict stability conditions near the wall, where the velocity is small. Therefore, implicit methods are usually applied, in this example the Crank–Nicholson method. The vertical velocity component is solved afterwards using what is in fact an implicit method for an ordinary

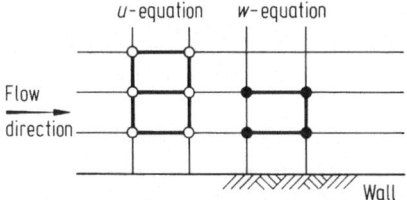

Fig. 14.4 Finite-difference "molecules" for boundary layers.

differential equation for w (see Fig. 14.4 for a schematic representation)

$$\frac{1}{2}\left(\frac{u_{k+1,j} - u_{k,j}}{\Delta x} + \frac{u_{k+1,j-1} - u_{k,j-1}}{\Delta x}\right) + \frac{w_{k+1,j} - w_{k+1,j-1}}{\Delta z} = 0 \, (14.20)$$

Because of the similarity with the problem of section 14.1, the same grid sizes are used, i.e. $\Delta z/a = 0.1$, $\Delta xv/a^2 u_0 = 0.05$ and $\theta = 0.5$ (keep in mind that the viscosity v plays the same role as the diffusion coefficient D in the transport equation). Starting with a linear profile, you see in Fig. 14.5 that the streamlines approach their equilibrium positions. The pressure gradient is expressed in a dimensionless form as $S' = ga^2(S - \rho^{-1} \partial p/\partial x)/vu_0$. This tends to the value 3 at equilibrium.

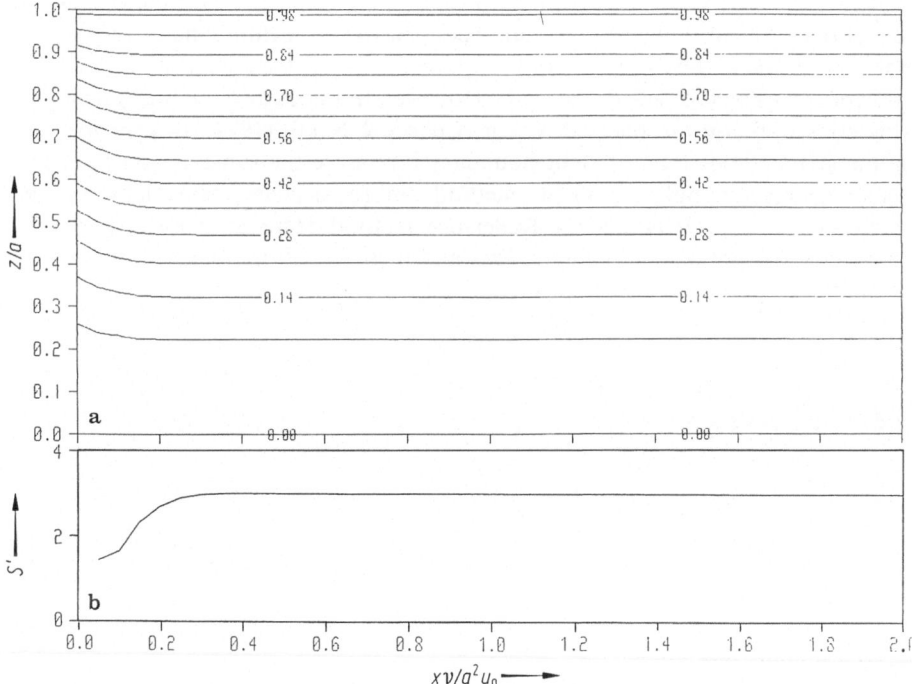

Fig. 14.5 Flow development in a river. a. Streamlines below which the indicated percentage of discharge flows. b. Dimensionless pressure gradient S'.

The flow turns out to have been fully developed at about $xv/a^2 u_0 = 0.3$. This is much faster than in the case of suspended sediment (section 14.1). The explanation is that the pressure gradient helps to redistribute the flow; such a mechanism does not exist for sediment transport.

In most practical cases, the flow is turbulent. This means that it experiences a much greater internal friction due to turbulent mixing. For flow in a river, theories on turbulence indicate that the effective or turbulent viscosity behaves like

$$v = v_0 \frac{z}{a}\left(1 - \frac{z}{a}\right) \tag{14.21}$$

You will have to adapt the no-slip boundary condition on the bottom as follows

$$u = 0 \quad \text{at } z = z_0 \tag{14.22}$$

This expresses the fact that the bottom has a certain roughness; the velocity goes to zero somewhere within this roughness range, of which z_0 is a measure; experimentally it is found to be roughly 3% of the average height of the roughness elements (sand grains, ripples, etc.). You may check that the turbulent velocity profile becomes

$$u(z) = \frac{g S a^2}{v_0} \ln \frac{z}{z_0} \tag{14.23}$$

which is the well-known logarithmic profile. Away from the bottom, it is much flatter than the parabolic profile found before; near the bottom it has a very strong velocity gradient. The latter feature requires a special treatment in a numerical method, by using the "law of the wall". However, this goes into too much detail for this book and you are referred to, e.g., Cebeci & Smith (1974). In principle, the numerical methods for turbulent boundary layers are the same as you have just seen. A very popular implicit method, more or less equivalent to the Crank–Nicholson method, is the Keller-box method. Here, all questions are split up into first-order differential equations and these are discretized within one grid interval. See again the book just mentioned.

14.6 Exercises

1. Numerical methods can be easily adapted to the form of eq. (14.3). As an example, write down the Crank–Nicholson method for it, without dividing by the flow velocity.
2. If you use the FTCS method for the numerical solution of eq. (14.3), what will be the stability conditions?
3. Do you agree with the grid sizes in the examples of sections 14.2 and 14.5?
4. What is the equivalent of the cell-Reynolds criterion in the case of boundary layers?

Chapter 15
Long Waves

15.1 Simplified Formulation

Tidal waves in rivers and seas, flood waves in rivers, but also oscillations in harbour basins are long waves relative to the water depth. A mathematical formulation can be obtained by integrating the general hydrodynamic equations over the depth or over a river cross-section. To understand the principles, it is sufficient to use a very much simplified set of equations in this chapter and the next one. For completeness, some of the corresponding results for the complete equations are given in appendix 1. For further reference, you can consult books such as Jansen (1979) or Cunge *et al.* (1980).

The first equation is the mass balance for water (4.2) which is simplified to

$$\frac{\partial h}{\partial t} + a\frac{\partial u}{\partial x} = 0 \tag{15.1}$$

The momentum equation says that the acceleration of a fluid particle is due to the water-level gradient and the bottom friction:

$$\frac{\partial u}{\partial t} + g\frac{\partial h}{\partial x} + ru = 0 \tag{15.2}$$

where

h = waterlevel relative to a horizontal datum level

a = water depth, considered approximately constant

r = a frictional coefficient (see app. 1)

u = flow velocity

Actually, h and u should be considered as small disturbances of an equilibrium situation (e.g. a steady flow). The equations (15.1) and (15.2) are known as the shallow-water, long-wave or Saint-Venant equations.

15.2 Characteristics

In Chapter 4, a characteristic was defined as a path in the x–t plane along which a certain quantity (there the concentration) is conserved. You could wonder, whether the long-wave equations have any comparable property. Beforehand, it is not known which quantity, if any, will be conserved, but it will most probably be a combination such as $f = mh + nu$ with as yet unknown coefficients m and n. Using the formulation of eqs. (15.1) and (15.2), multiplying them by m and n respectively and adding them, you find

$$m\frac{\partial h}{\partial t} + ng\frac{\partial h}{\partial x} + n\frac{\partial u}{\partial t} + ma\frac{\partial u}{\partial x} + nru = 0 \tag{15.3}$$

Now, in order to have the quantity f conserved, you should find something of the form

$$\frac{\partial f}{\partial t} + c\frac{\partial f}{\partial x} + \ldots = 0$$

or

$$m\frac{\partial h}{\partial t} + cm\frac{\partial h}{\partial x} + n\frac{\partial u}{\partial t} + cn\frac{\partial u}{\partial x} + \ldots = 0 \tag{15.4}$$

in which c is the celerity (unknown as well) of the characteristic. Comparing eqs. (15.3) and (15.4) you find this to be possible if

$$ng = cm, \qquad ma = cn$$

Evidently, this is possible only if

$$c = \pm(ga)^{1/2} \tag{15.5}$$

and

$$m = gn/c = \pm(g/a)^{1/2}\, n \tag{15.6}$$

The value of n is arbitrary and can be chosen to be unity. You find that there are two characteristics in each point. You can imagine that this is caused by the fact that the flow is described by two equations, so by a second order system of differential equations. If such a system has two separate characteristics, it is called hyperbolic; this usually means that you have a wave model. Along each characteristic, you have

$$\frac{df}{dt} = \frac{\partial f}{\partial t} + c\frac{\partial f}{\partial x} = -ru \tag{15.7}$$

indicating that the quantity f is not really conserved, but it is damped by bottom friction (compare with the damping of concentration by decay in Chapter 4). In the absence of bottom friction, the quantities f are conserved; they are then called the

Riemann invariants:

$$f = u + (g/a)^{1/2}h \qquad \text{if } c = +(ga)^{1/2} \tag{15.8}$$

$$F = u - (g/a)^{1/2}h \qquad \text{if } c = -(ga)^{1/2} \tag{15.9}$$

So keep in mind that f is related to the positive (forward propagating) characteristic line, and F to the negative one. A characteristic line is briefly called a characteristic. It has something to do with a wave running in that direction, which you can understand as follows (Fig. 15.1).

Suppose you have a stretch of a river between two boundaries. If you want to know what is happening at a point A, you could use the two characteristics AB and AC. For simplicity forget the bottom friction for a while. Then

along AB: $\quad f = u + (g/a)^{1/2}h = \text{known}$

along AC: $\quad F = u - (g/a)^{1/2}h = \text{known}$

provided you would know these values at time $t = 0$ (in B and C). Then you could solve h and u in point A. Apparently, you should know both h and u at the initial time: you need two initial conditions in each point of the river reach (as the points B and C may be anywhere). You may now check what happens if you start with a "hump" in the waterlevel at time zero; it will fall apart into two smaller humps travelling in either direction at the characteristic speed. In section 15.4, you will see that bottom friction complicates this picture.

The construction works for all points within the triangle DEF, but what about point I, for example? Here everything could be computed in the same way if the point G is known, so the problem is shifted towards the boundaries of the region. At G, you have only one characteristic GB, so F is known. However, this is only one equation with two unknowns, so you need one more equation: you need one boundary condition. This may be either a value for h or for u, or a combination of both.

It will now be clear that you can reach all points in the region at any time by repeated application of the characteristic relations. If bottom friction is not neglected, the quantities f and F are no longer constant along characteristics. The construction gets more complicated then, but the principle remains the same.

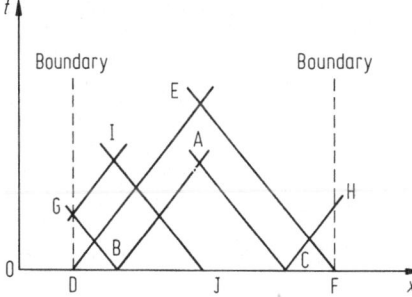

Fig. 15.1 Characteristics and boundary conditions.

From this example, you find what turns out to be a general rule for hyperbolic equations: the number of initial or boundary conditions needed is equal to the number of characteristics entering the region at that particular point. Check for yourself that this indeed describes what you just found.

Some examples:

(i) at the mouth of a river, the waterlevel can be specified as a function of time, describing the tidal fluctuation of the sea water level;
(ii) at an upstream boundary, the river discharge can be specified, which is a combination of velocity and depth (which depends on the waterlevel).
(iii) at a downstream weir, the flow over the weir depends on the local waterlevel; this gives again a "mixed" boundary condition.

15.3 Weakly Reflecting Boundary Condition

Suppose that you choose the upstream boundary of a tidal-river model so close that there is still a tidal influence there; however, not knowing this, you impose a constant discharge as a boundary condition. What will happen then is that the tidal wave will reflect at that boundary. You can understand this by realizing what a constant-discharge condition actually means: a dead end of the river with a pumping station discharging the specified amount into the river.

There is only one good solution to this problem: shifting the boundary to a location more upstream where you are sure that no tidal fluctuations exist anymore. How to estimate this location is treated in the next section. However, it may be useful to keep the boundary where it is (in order to limit the computational effort) and to use a boundary condition that is non-reflecting, or at least weakly reflecting.

This can be obtained using the theory of characteristics. It is assumed that outside the region the river is extended by a fictitious river with the same dimensions but no friction (Fig. 15.2). Then in a point A on the boundary, you

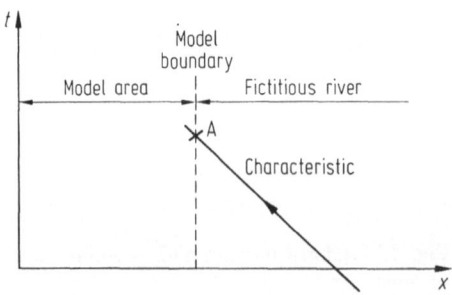

Fig. 15.2 Weakly reflecting boundary condition.

know the value of F if it is known at "infinity", i.e. very far upstream:

$$u - (g/a)^{1/2}h = u_\infty - (g/a)^{1/2}h_\infty \tag{15.10}$$

For example, at infinity there is an equilibrium flow, which means that the deviations h and u are zero. Equation (15.10) can be used as a boundary condition at A and it works as a weakly reflecting condition. There remains some reflection as you have introduced a "jump" in the bottom friction at the boundary.

15.4 Example

Consider a river with depth $a = 7$ m, base discharge $Q = 500$ m^3/s, frictional coefficient $c_f = 0.006$ and a tidal amplitude at the mouth of 1 m. It can be shown with the method of the next section that the tidal wave will penetrate the river about 100 km. In Fig. 15.3 the numerical results are shown of numerical models which are identical except for the upstream boundary:

(a) boundary at 155 km, boundary condition: $Q = 500$.
(b) boundary at 55 km, same condition.
(c) boundary at 55 km, eq. (15.10) as boundary condition.

In Fig. 15.3a, you see that the tidal wave is indeed very small at $x = 100$ km. Figure 15.3b shows the strong reflections caused by a fixed discharge. Finally, Figure 15.3c indicates that a weakly reflecting boundary condition gives results that, though not extremely accurate, have some significance.

15.5 Wave Propagation

Using the approach of sinusoidal waves that you have seen before, it is possible to get information complementary to the theory of characteristics. To do this correctly, you have to use the complete equations in linearized form instead of eqs. (15.1) and (15.2). This makes the analysis somewhat involved and it is moved to Appendix 1. The idea is to try a solution of the form

$$h(x, t) = H \exp(\lambda x - i\omega t)$$
$$u(x, t) = U \exp(\lambda x - i\omega t) \tag{15.11}$$

where λ, H, U are complex quantities. Substituting this into the equations, a condition for λ is found and the (complex) ratio of H and U can be determined. For the details, see Appendix 1.

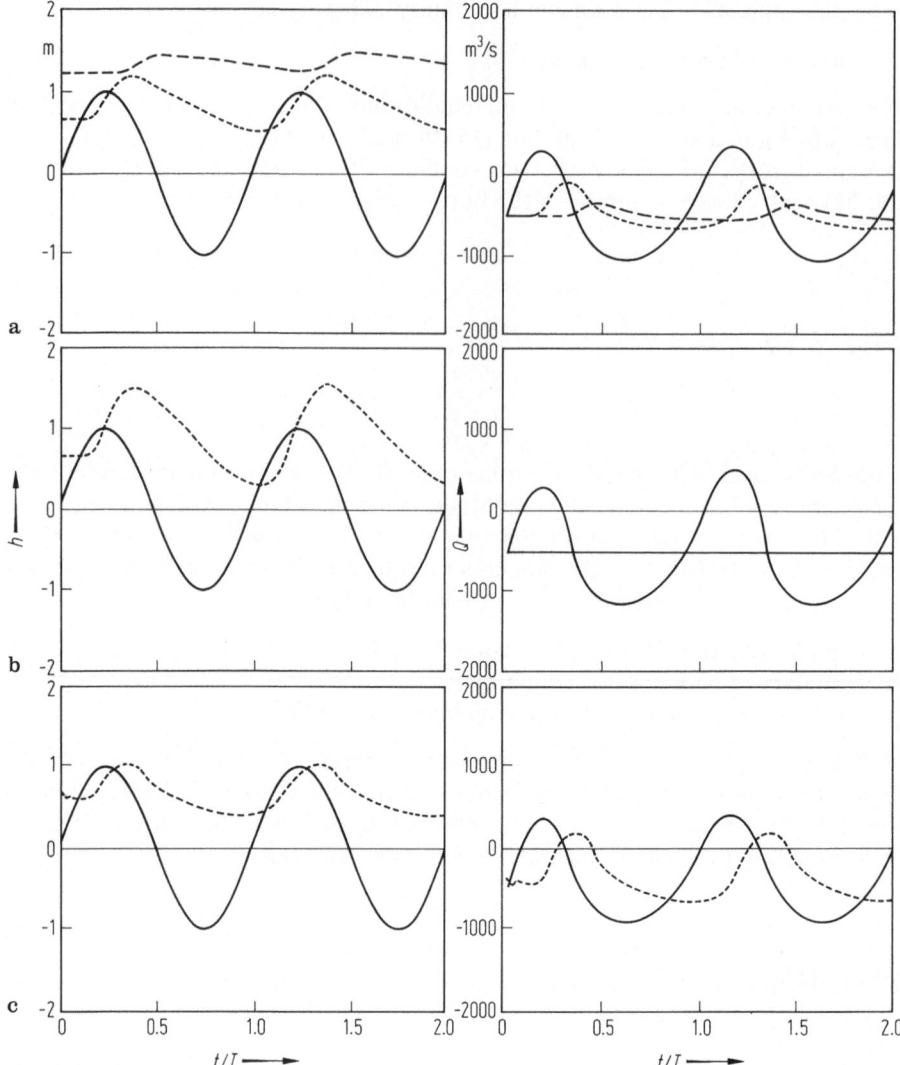

Fig. 15.3 (Left) Water levels at $x = 0$, 50, 100 km from the river mouth; (right) discharge at $x = 5$, 55, 105 km from the river mouth. Cases a, b and c explained in the text.

If the boundary condition for the water level is applied at the river mouth and the positive x-axis is directed up the river, only the root with negative real part can be used, as the other increases exponentially (the river is assumed infinitely long). Writing $\lambda = -\mu + ik$, you get

$$h(x, t) = H \exp(-\mu x) \exp\{i(kx - \omega t)\} \tag{15.12}$$

which describes a damped wave travelling at a speed $c = \omega/k$ (similarly for u).

Figure 15.4 shows in a dimensionless form the wave speed

$$c' = c(ga)^{-1/2} \tag{15.13}$$

and Fig. 15.5 the dimensionless relaxation or damping length (amplitude decay to e^{-1})

$$L' = \omega/\{2\pi\mu(ga)^{1/2}\} \tag{15.14}$$

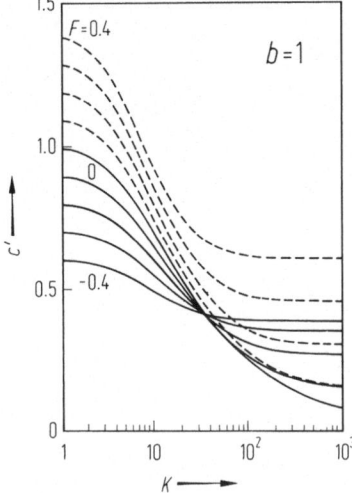

Fig. 15.4 Dimensionless wave speed.

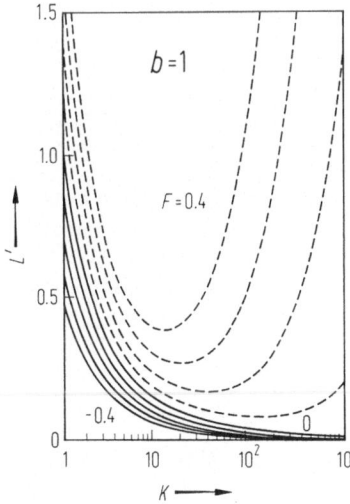

Fig. 15.5 Dimensionless damping length.

Moreover, Fig. 15.6 shows the dimensionless amplitude of the velocity

$$U' = U/H(a/g)^{1/2} \tag{15.15}$$

and Fig. 15.7 the phase angle between velocity and waterlevel

$$\phi = \arg(U/H) \tag{15.16}$$

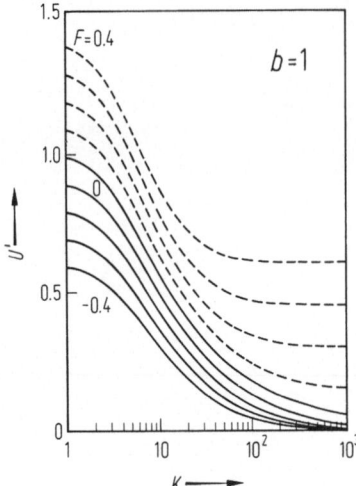

Fig. 15.6 Dimensionless amplitude of velocity.

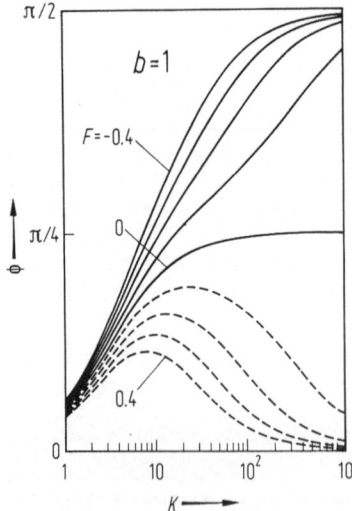

Fig. 15.7 Phase angle between velocity and water level.

All quantities are functions of the Froude number $F = u_0/(ga)^{1/2}$ and what is called (for lack of a better name) the friction parameter

$$K = r/\omega = u_0 T c_f/a \qquad (15.17)$$

in which u_0 is the mean velocity or (if this is larger) a fraction of the oscillating velocity (see again appendix 1). Anyway, K indicates the importance of bottom friction and the wave period.

For small values of K and F, you observe from the figures that the wave speed approaches the characteristic speed, there is no wave damping (relaxation length very large), the velocity variation is in phase with the waterlevel variation and its amplitude agrees with the characteristic theory, i.e. you are back in section 15.2. However, for larger values of K, quite drastic changes can occur: much smaller wave speeds, significant damping, phase shifts and decrease of the velocity amplitude. Therefore, you should not be surprised if you notice such phenomena in a numerical simulation. Actually, using this theory, you can even check your numerical results.

15.6 Example

Consider a tidal river with depth 7 m and a tidal amplitude of 0.5 m at the river mouth. The friction coefficient is $c_f = 0.004$. The width of the river is 500 m; the river discharge is 1000 m^3/s. For a semidiurnal tide with wave period 12.5 h, you find $K = 6.5$ and $F \simeq 0$. From Figs. 15.4 to 15.7, you read

$$c' = 0.77 \quad \text{so} \qquad c = 6.4 \text{ m/s}$$
$$L' = 0.20 \qquad\qquad L = 75 \text{ km}$$
$$U' = 0.64 \qquad\qquad U = 0.38 \text{ m/s}$$
$$\phi = 0.55 \text{ rad} \qquad \text{phase shift} = 0.55 * 12.5/2\pi = 1.13 \text{ h}$$

The results of a numerical simulation with the complete (non-linear) equations are shown in Fig. 15.8 for comparison. You may now check the previous theoretical estimates with the "exact" results.

The travel time over a distance of 50 km would be 2.15 h according to the estimate. From Fig. 15.8, you read the following times:

peak of h from 0 km to 50 km	1.5 h
trough of h (same)	2.75 h
peak of Q from 5 to 55 km	1.75 h
minimum of Q (same)	2.75 h

On the average, this agrees well with the estimate. You observe the nonlinear behaviour in the different travel times for peaks and troughs (the characteristic speed varies with the water depth).

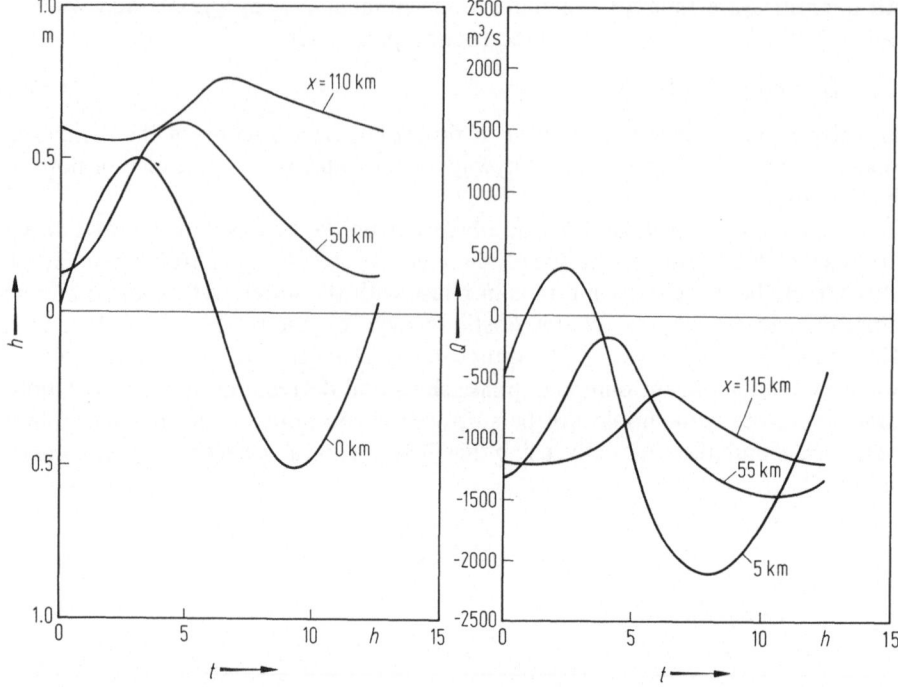

Fig. 15.7 Phase angle between velocity and water level.

Secondly, the amplitude damping is considered. You read from the figure:

	h			Q		
	0	50	110	5	55	115 km
max	0.5	0.62	0.76	400	− 180	− 610
min	− 0.5	0.13	0.55	− 2070	− 1430	− 1200
ampl.	0.5	0.25	0.11	1235	625	295
damping factor		0.450	0.22		0.51	0.24

From the estimate, at a distance of 50 km, you find a damping factor of exp (
$- x/L) = 0.51$ and at 110 km, it will be 0.23. Both agree very well with the
numerical values.

Thirdly, the estimated amplitude of discharge at the river mouth is 0.38 ∗ 3500
$= 1340 \, m^3/s$; the numerical value is 1235, which is again quite close.

Finally, the phase shift between discharge (or velocity) and water level can be
read from the times of peaks and troughs, e.g. at the river mouth:

  max Q − peak h: 1 h
  min Q − trough h: 1.5 h.

The estimate is 1.13 h, which agrees well. It can be concluded that the linear analysis discussed in this section gives quite useful estimates to predict what will be happening and to check the numerical results.

15.7 Exercises

1. Determine the characteristics for the diffusion equation (7.1) + (7.2). Use the fact that h and a have the same time rate of change. Assume a in the diffusion coefficient to be constant. Temporarily add a term $\varepsilon \, \partial u / \partial t$ to eq. (7.2) and take $\varepsilon = 0$ at the end. You will find that there are two coincident characteristics with infinite celerity; this is called the parabolic case.
2. For the example of section 15.4, where would you locate the upstream boundary of the numerical model such that the tidal amplitude is negligible there?

Chapter 16
Numerical Methods for Long Waves

16.1 Leap-Frog Method

From the simplified Saint–Venant equations (15.1) and (15.2), you can easily eliminate one of the unknowns, so that you obtain one differential equation with one unknown. However, for the general equations given in appendix 1, this is much more complicated and moreover not very useful. The straightforward way is to discretize the equations directly. To this end, a grid in the x–t plane is chosen with spatial grid size Δx and time step Δt (Fig. 16.1). If you use central differences both in space and time, the leap-frog method results (see also section 5.2):

$$\frac{h_j^{n+1} - h_j^{n-1}}{2\Delta t} + a\frac{u_{j+1}^n - u_{j-1}^n}{2\Delta x} = 0 \tag{16.1}$$

$$\frac{u_{j-1}^{n+2} - u_{j-1}^n}{2\Delta t} + g\frac{h_j^{n+1} - h_{j-2}^{n+1}}{2\Delta x} + \frac{1}{2}r(u_{j-1}^{n+2} + u_{j-1}^n) = 0 \tag{16.2}$$

If the values at times n and $n-1$ are known, you can solve the values at time $n+1$ and $n+2$. You will note that to compute h_j^{n+1}, you need u-values only at points $j-1$ and $j+1$, and conversely (Fig. 16.1). This gives a leap-frog type pattern of grid points. It turns out that you do not need both h and u values at all grid points, but only one of them, and even nothing in half the number of points. This is called a staggered grid. Yet, the effective grid size remains Δx, Δt. Consequently, the method is quite efficient: the solution is obtained at a certain accuracy with a small number of computational operations.

Actually, you could apply this method for each variable at each grid point, but you would get four independent grids which do not influence one another.

There are a few problems with this method:

(i) As in section 5.2, you need two time levels to start the method. Therefore, for the first time step you have to apply a different (one-step) method.
(ii) Boundary conditions for either h or u can be accommodated easily by taking care that the boundary grid points are of the correct type (i.e. h or u). However,

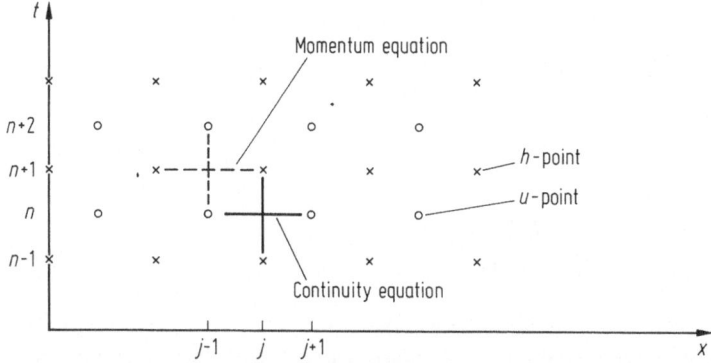

Fig. 16.1 Leap-frog method with grid staggered in space and time.

a mixed condition is less simple. Some sort of averaging will be needed, but this is not elaborated here.

(iii) It is not simple to include the convective terms as this would require u_j^{n+1} and $u_{j-\frac{1}{2}}^{n+1}$ in eq. (16.2). Neither are available in the grid. To keep the method explicit, these terms would have to be evaluated at time level n, which destroys the second order accuracy. This can be done only if the convective terms are not very important (which, fortunately, is sometimes the case). For this reason, the method is not usually applied in general purpose programs.

In an exercise, it is shown that the leap-frog method has second-order accuracy both in space and time.

16.2 Stability of the Leap-Frog Method

The Von Neumann method can be used again for the analysis of stability. In this case, both variables are assumed to behave as a harmonic function:

$$h_j^n = H^n \exp{(ij\xi)} \tag{16.3}$$

$$u_j^n = U^n \exp{(ij\xi)} \tag{16.4}$$

where H^n and U^n are amplitudes at time level n (note that n is not an exponent but a superscript here). For ease of analysis, you assume the finite difference equations to be valid at all grid points (which is correct although you will in practice use only one quarter of them). From eqs. (16.1) and (16.2) (the latter now at time level $n + 1$), you find by substitution

$$H^{n+1} - H^{n-1} + 2a\frac{\Delta t}{\Delta x}U^n i \sin \xi = 0 \tag{16.5}$$

$$U^{n+1} - U^{n-1} + 2g\frac{\Delta t}{\Delta x}H^n i \sin \xi + r\Delta t(U^{n+1} + U^{n-1}) = 0 \tag{16.6}$$

which allows solutions of the form

$$U^n = \rho^n U, \qquad H^n = \rho^n H$$

where H, U are the initial amplitudes. Note that n in ρ^n is an exponent. Substitution into eqs. (16.5) and (16.6) gives a system of two linear, homogeneous equations in H, U, which has a nonzero solution only if the determinant of coefficients is zero. This results in a quadratic equation for ρ^2:

$$(1 + r\Delta t)\rho^4 + 2(2\sigma^2 \sin^2 \xi - 1)\rho^2 + 1 - r\Delta t = 0 \tag{16.7}$$

with two roots

$$\rho_{1,2}^2 = [1 - 2\sigma^2 \sin^2 \xi \pm \{(1 - 2\sigma^2 \sin^2 \xi)^2 - (1 - r^2\Delta t^2)\}^{1/2}]/(1 + r\Delta t) \tag{16.8}$$

Here $\sigma = (ga)^{1/2} \Delta t/\Delta x$ is the Courant number; you see that the characteristic speed occurs in it.

In fact there are now four roots for ρ, two of which correspond to the two physical waves. The other two (the negative square roots from the right-hand side of eq. (16.8)), are spurious: they are purely of a numerical origin. The reason is that the finite-difference equations are of higher order than the differential equations due to the use of central differences in time. If the spurious roots do not exceed unity (see below) they are not particularly harmful if a staggered grid is used. The spurious waves have an oscillating character, as you can see from eq. (16.8) if $\sin \xi \simeq 0$, resulting in $\rho \simeq -1$. However, on a staggered grid such waves cannot be represented. The two significant roots of eq. (16.8) can be considered as propagation factors over a double time step (as this involves ρ^2).

For the case without bottom friction, eq. (16.7) becomes

$$\rho^4 + 2\rho^2(2\sigma^2 \sin^2 \xi - 1) + 1 = 0 \tag{16.9}$$

From this, you can conclude under which conditions the roots do not exceed unity, even without solving the two roots. This type of reasoning is very common in stability considerations. First of all, as the coefficients are real, the roots are either real or conjugate complex. The product of the roots is 1 (the last coefficient). If the roots were real, one of them would therefore exceed unity and the method would be unstable. Therefore, the roots should be complex, i.e. the discriminant should be negative:

$$(2\sigma^2 \sin^2 \xi - 1)^2 - 1 \leqslant 0$$

or

$$-1 < 2\sigma^2 \sin^2 \xi - 1 < 1$$

As this must be valid for the most unfavourable value of ξ, you find as a stability condition

$$|\sigma| \leqslant 1 \tag{16.10}$$

as usual for explicit methods.

The explanation of this criterion is exactly the same as in section 5.3 and applies to both characteristics. Note that even in case of stability, the amplification factors ρ are exactly unity in absolute value, so the method is only marginally stable. There is no wave amplification, but no attenuation either. This is different of course if friction is present, as you can see from eq. (16.7).

16.3 Example

In Fig. 16.2, an elongated harbour basin is shown. If waves of a certain period enter from the sea, the basin may get into resonance: a standing wave may develop. To study the sensitivity, a simple numerical model is set up using the leap-frog method. The grid layout is also shown in Fig. 16.2. The total length of the basin is 10 km; the grid size of 769 m is chosen such that a waterlevel point is at the seaward boundary and a velocity-point at the other end. This corresponds to the boundary conditions; waterlevel as a function of time at the seaward end; velocity zero at the landward end. The depth is 10 m. The resonant wave has a length of four times the basin length; its period is approximately $T = 3260$ s. A time step $\Delta t = 54$ s has been used, which gives a Courant number $\sigma = 0.85$. This is on the safe side of the stability limit and allows some variation of the wave speed. At the boundary, a periodic waterlevel variation with amplitude 0.10 m has been applied.

Three cases have been considered:

	(a)	(b)	(c)
T	3260	3260	4075
c_f	0.004	0·005	0·004

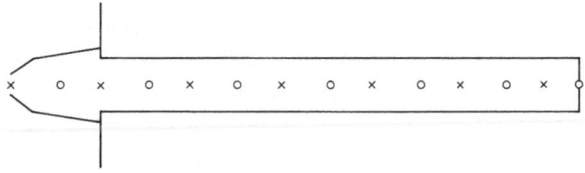

Fig. 16.2 Harbour basin with numerical grid.

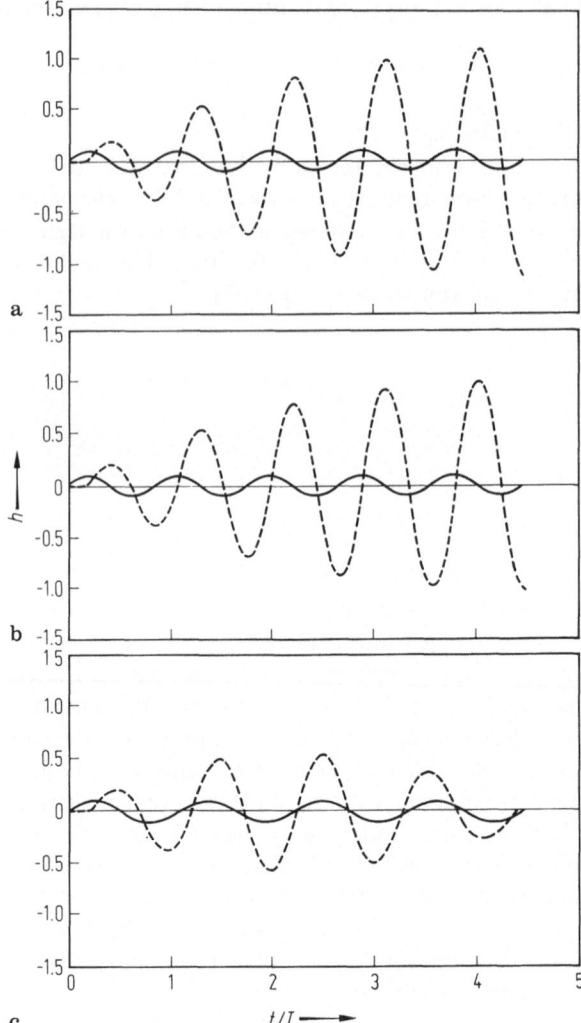

Fig. 16.3 Computed water levels in harbour basin.

The computed waterlevel variation at the landward end for the three cases is shown in Fig. 16.3, together with the boundary condition. Case (a) shows that there is indeed a strong amplification; after five wave periods the amplitude is even still growing. Case (b) with a greater bottom friction shows the same picture with a slightly smaller amplitude. The "equilibrium" amplitude is actually determined by bottom friction: without it, you would get an infinite amplitude. The results of case (c) with a 25% increase in the wave period, appear to be odd: no equilibrium wave is found. However, with a little thought the picture can be understood. The boundary condition introduces a wave period of $1.25\,T$ if T is the resonant wave period of 3260 s. The boundary condition is started impulsively at time $t = 0$; this introduces a transient in the basin which will certainly contain the resonant wave. Therefore, you get a mixture of the two waves, roughly as follows (the amplitudes are not so important for the purpose of explanation):

$$h = \sin 2\pi t/T + \sin 2\pi t/(1.25\ T)$$

$$= 2 \sin 0.9\ 2\pi t/T \sin 0.1\ 2\pi t/T$$

This describes a wave with wave period $T/0.9$ with an amplitude modulation: the amplitude varies slowly with a period of $10T$. Inspection of Fig. 16.3c confirms this.

Unfortunately, you have now a result that is not useful physically, because the transient wave is "trapped" by a fixed boundary condition. This is a perfect case to use a weakly reflecting boundary condition, which allows the transient wave to escape. To do this on the staggered grid of Fig. 16.2 is a little involved. The idea is to use the continuity equation at the boundary point with a forward space difference for the velocity gradient. The value of the velocity at the boundary point is not available, but it can be eliminated using the characteristic relation 15.10. At infinity, a travelling wave with amplitude 0.10 m is assumed. With this condition, Fig. 16.4 gives results for cases (a) and (c) (note the change of scale). An equilibrium condition develops much quicker now, but also the amplification of the wave is

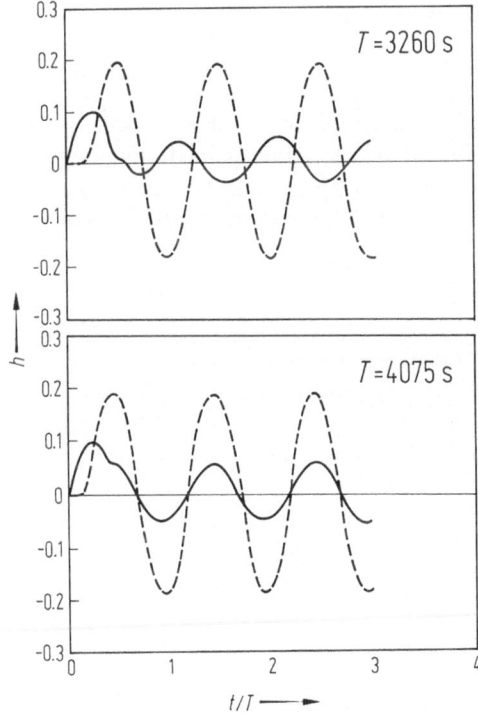

Fig. 16.4 Computed water levels using weakly reflecting boundary condition.

different. Numerically, the following values are found for the ratio between amplitudes at the landward and seaward ends of the basin:

T	fixed boundary condition	weakly reflective boundary condition
3260	> 10	3.8
4075	$\simeq 5$	3.2

The fixed-boundary results turn out to give a false picture of the resonance effect.

16.4 Implicit Methods

There are, of course, more explicit finite-difference methods for the long-wave equations than leap-frog, but for reasons explained in the next section, they are not very popular. Implicit methods are often applied, such as the Crank–Nicholson method. You will recognize the structure of the equations:

$$\frac{h_j^{n+1} - h_j^n}{\Delta t} + \theta a \frac{u_{j+1}^{n+1} - u_{j-1}^{n+1}}{2\Delta x} + (1-\theta)a\frac{u_{j+1}^n - u_{j-1}^n}{2\Delta x} = 0 \qquad (16.11)$$

$$\frac{u_{j-1}^{n+1} - u_{j-1}^n}{\Delta t} + \theta g \frac{h_j^{n+1} - h_{j-2}^{n+1}}{2\Delta x} + (1-\theta)g\frac{h_j^n - h_{j-2}^n}{2\Delta x} +$$

$$+ r\{\theta u_{j-1}^{n+1} + (1-\theta)u_{j-1}^n\} = 0 \qquad (16.12)$$

In this case, staggering of the grid in time is not possible, but it is in space, so you can use a grid as shown in Fig. 16.5. A second grid could be defined, shifted by one grid size to the left or right, so you are now solving for one half of the possible variables. The grid should again be chosen such that the correct type of points are on the boundaries.

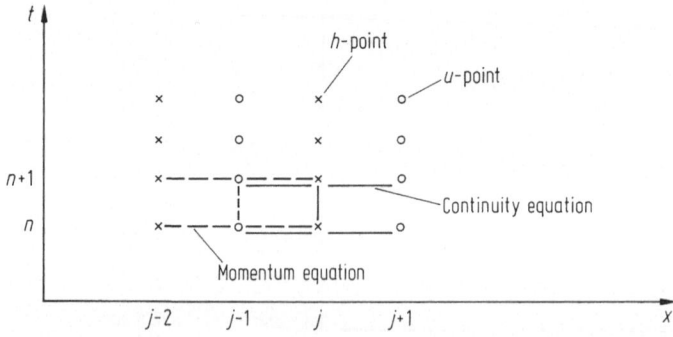

Fig. 16.5 Space-staggered grid for the Crank–Nicholson method.

If you consider eqs. (16.11) and (16.12) as a system of linear equations, the matrix turns out to be of the tridiagonal type, which can be solved efficiently using the Thomas algorithm (section 7.4).

In practice, the complete equations (Appendix 1) will be used. For the convective terms, you will need u (or Q) values in h-points. These can only be obtained by interpolation:

$$2\Delta x \frac{\partial}{\partial x}(Q^2/A_s) \approx (Q^2/A_s)_{j+1} - (Q^2/A_s)_{j-1} = \tfrac{1}{2}\{(Q^2/A_s)_{j+2} + (Q^2/A_s)_j\} +$$

$$-\tfrac{1}{2}\{(Q^2/A_s)_j + (Q^2/A_s)_{j-2}\} = \tfrac{1}{2}(Q^2/A_s)_{j+2} - \tfrac{1}{2}(Q^2/A_s)_{j-2} \qquad (16.13)$$

The conclusion is that you are effectively using the double grid size for these terms. This is, again, acceptable if the convective terms are not too important. If they are, you will have to reduce the grid size for accuracy and you will lose the advantage of the staggered grid. Moreover, in this case the matrix gets pentadiagonal and the algorithm more complicated. For a more extensive discussion see Cunge et al. (1980).

Another difficulty using the complete equations is that they are nonlinear. They are linearized in each time step and iterated if it is necessary (it often is not). See again Cunge et al. (1980).

A variant of the Crank–Nicholson method is obtained by applying differences over only one grid interval. The resulting method is known by the name of Preissmann (also Wendroff; the method seems to have been invented by various people simultaneously) or as the four-point method. The layout of grid points is shown in Fig. 16.6; now both variables are defined in each grid point, so there is no staggering. You might think that there is now only first-order accuracy by using forward or backward differences, but in fact the differences are central with respect to the centre of the grid interval (i.e. location $(j-\tfrac{1}{2})\Delta x$). For the time difference, some average is needed, so you get

$$\frac{1}{2}\left(\frac{h_j^{n+1} - h_j^n}{\Delta t} + \frac{h_{j-1}^{n+1} - h_{j-1}^n}{\Delta t}\right) + \theta a \frac{u_j^{n+1} - u_{j-1}^{n+1}}{\Delta x} + (1-\theta)a\frac{u_j^n - u_{j-1}^n}{\Delta x} = 0$$

and similarly for the momentum equation.

$$(16.14)$$

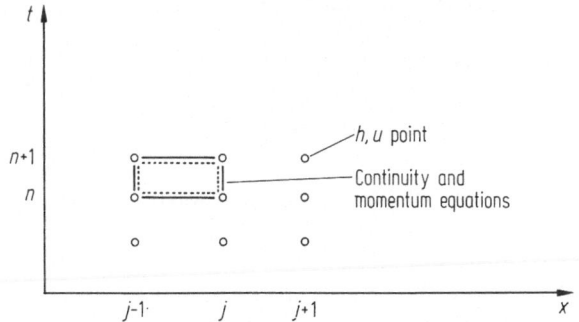

Fig. 16.6 Grid for the Preissmann method.

If you order the equations, you obtain a system with a pentadiagonal matrix. However, it has a special structure:

$$a_k h_k + b_k Q_k + c_k h_{k+1} + \qquad d_k Q_{k+1} \qquad\qquad = \dots$$
$$A_k h_k + B_k Q_k + C_k h_{k+1} + \qquad D_k Q_{k+1} \qquad\qquad = \dots$$
$$a_{k+1} h_{k+1} + b_{k+1} Q_{k+1} + c_{k+1} h_{k+2} + d_{k+1} Q_{k+2} = \dots$$
$$A_{k+1} h_{k+1} + B_{k+1} Q_{k+1} + C_{k+1} h_{k+2} + D_{k+1} Q_{k+2} = \dots$$

This allows a straightforward extension of the Thomas algorithm (Cunge *et al.* 1980). You may check (by substitution) that the following holds:

$$Q_k = F_k h_k + G_k \tag{16.15}$$

$$h_k = f_k h_{k+1} + g_k Q_{k+1} + p_k \tag{16.16}$$

and from the same substitution you can figure out the recurrence relations for the coefficients f, g, p, F, G. The forward sweep now consists of computing these coefficients; the backward sweep solves h and u from eqs. (16.15) and (16.16) starting at the end.

As there is no staggered grid, the Preissmann method is not particularly efficient, but it has one important advantage: the grid size can be changed very easily from one region to another. If you want to have a smaller grid size in a subregion or if you want to have your grid points at certain locations (e.g. where measurements are available), this can be done without disturbing the structure of the equations. In the other methods discussed above, a variation of grid size is very awkward.

In an exercise, the amplification factors for the two implicit methods are found to be:

$$\rho_{1,2} = \frac{1 + (1-\theta)x_{1,2}}{1 - \theta x_{1,2}} \tag{16.17}$$

with

$$x_{1,2} = -\tfrac{1}{2} r \Delta t \pm (\tfrac{1}{4} r^2 \Delta t^2 - \sigma^2 \sin^2 \xi)^{1/2} \tag{16.18}$$

for Crank–Nicholson, and

$$x_{1,2} = -\tfrac{1}{2} r \Delta t \pm (\tfrac{1}{4} r^2 \Delta t^2 - 4\sigma^2 \tan^2 \tfrac{1}{2} \xi)^{1/2} \tag{16.19}$$

for Preissmann. For the case without bottom friction ($r = 0$), it is easy to show that both methods are unconditionally stable (i.e. there are no restrictions on the Courant number, compare section 7.3) provided that you choose the factor θ in the interval $\tfrac{1}{2} \leqslant \theta \leqslant 1$ as before. The physical explanation of this is (as in section 7.3) that, due to the implicit character, the full dependence region for each point is taken into account, whatever the time step is (Fig. 16.7).

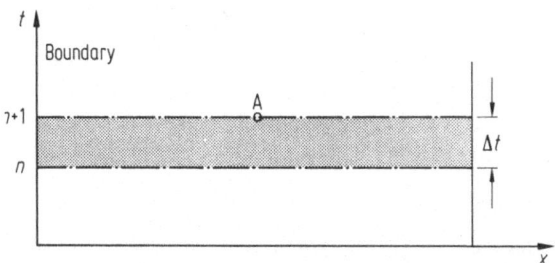

Fig. 16.7 Numerical dependence region for implicit methods.

16.5 Numerical Wave Propagation

In order to apply any of the numerical methods of this chapter to a practical problem, you will have to make a choice of the particular method and of the grid size and time step. An important factor is the accuracy with which long waves are reproduced numerically. There is an experimental way of judging this: do the computation twice, with different grid size and/or time step. If you find different results you have a warning that something is wrong and you will have to refine the grid even further. If the results from two runs with different numerical parameters agree, you have at least an indication that they are correct. Still, in the latter case it is important to have an idea whether the two results might be equally far from the truth.

An estimate can be made beforehand using the method of linear wave propagation as in many other places in this book. To keep things simple, consider the basic equations (16.1) and (16.2) without bottom friction $r = 0$. It is possible to do the analysis with bottom friction included, but it gets quite complicated. For convenience, write the equations in vector form:

$$\frac{\partial \mathbf{v}}{\partial t} + A \frac{\partial \mathbf{v}}{\partial x} = 0 \tag{16.20}$$

with

$$\mathbf{v} = \begin{pmatrix} h \\ u \end{pmatrix} \qquad A = \begin{pmatrix} 0 & a \\ g & 0 \end{pmatrix}$$

The solution has the form of propagating waves

$$\mathbf{v} = \mathbf{v}_1 \exp\left(ikx - i\omega t\right) \tag{16.21}$$

The value of ω is found by substituting (16.21) into eq. (16.20). This results in a linear homogeneous system of equations for $\mathbf{v}_1$, which has a nonzero solution only if the determinant is zero:

$$\det\left(-\omega I + kA\right) = 0$$

which gives

$$\omega = \pm k(ga)^{1/2} = kc$$

where c is the speed of propagation. So far nothing new; however, for each value of c, which actually is an eigenvalue of A, an eigenvector can be defined by using one of the two equations in (16.20):

$$-\omega h_1 + kau_1 = 0$$

This means that the eigenvectors are

$$\mathbf{v}_{1,2} = \begin{pmatrix} 1 \\ c/a \end{pmatrix} \quad \text{for} \quad c_{1,2} = \pm(ga)^{1/2} \tag{16.22}$$

The general solution with wave number k will now be a linear combination of these two solutions:

$$\mathbf{v} = \alpha_1 \mathbf{v}_1 \exp\{ik(x-ct)\} + \alpha_2 \mathbf{v}_2 \exp\{ik(x+ct)\} \tag{16.23}$$

with the coefficients $\alpha_{1,2}$ determined by the initial conditions. In general, you get two waves propagating into opposite directions. If the initial conditions are such that, e.g., $\alpha_2 = 0$, the remaining component behaves like

$$\frac{\partial \mathbf{v}}{\partial t} + c_1 \frac{\partial \mathbf{v}}{\partial x} = 0$$

which represents a simple wave. The other component, if present, behaves in the same way, but propagates in the opposite direction.

The system (16.20) is now approximated using one of the numerical methods, say, the Crank–Nicholson method. In vector form, this reads

$$\frac{\mathbf{v}_j^{n+1} - \mathbf{v}_j^n}{\Delta t} + \theta A \frac{\mathbf{v}_{j+1}^{n+1} - \mathbf{v}_{j-1}^{n+1}}{2\Delta x} + (1-\theta)A \frac{\mathbf{v}_{j+1}^n - \mathbf{v}_{j-1}^n}{2\Delta x} = 0 \tag{16.24}$$

Assume this to have solutions of the form

$$\mathbf{v} = \mathbf{v}_1 \rho^n \exp(ikx)$$

and substitute, then you find

$$\left[(\rho - 1)I + A \frac{\Delta t}{\Delta x}(\theta\rho + 1 - \theta)i\sin\xi \right] \mathbf{v}_1 = 0$$

which is again a linear and homogeneous system, of which the determinant must be zero. This is another eigenvalue problem, involving the same eigenvectors as above (viz. those of the matrix A), but with eigenvalues satisfying

$$(\rho - 1) + \sigma_{1,2}(\theta\rho + 1 - \theta)i\sin\xi = 0$$

or

$$\rho_{1,2} = \frac{1-(1-\theta)\sigma_{1,2}i\sin\xi}{1+\theta\sigma_{1,2}i\sin\xi} \tag{16.25}$$

As you see, there is an amplification factor corresponding to each value of the propagation (or characteristic) speed. The Courant number $\sigma = (ga)^{1/2}\Delta t/\Delta x$ is again involved as you already suspected.

The general numerical solution for wave number k is now

$$\mathbf{v} = (\alpha_1 \mathbf{v}_1 \rho_1^n + \alpha_2 \mathbf{v}_2 \rho_2^n)\exp(ikx) \tag{16.26}$$

with the same α values as above, because these are determined by decomposing the initial condition into the two eigenvectors. Comparing this with eq. (16.23), you conclude that each wave can be considered separately with its damping rate

$$d(n) = |\rho|^n$$

and relative phase speed

$$c_r = -\arg(\rho)/(2\pi\Delta t/T)$$

(compare section 5.5). The entire reasoning applies to the other finite-difference methods as well, with the only difference that they have their own amplification factors, i.e.

leap-frog $\qquad\qquad \rho = -\sigma i \sin\xi \pm (1-\sigma^2\sin^2\xi)^{1/2}$ $\qquad\qquad$ (16.27)

(with the minus sign for the spurious or parasitic wave)

Preissmann $\qquad\qquad \rho = \dfrac{1 - 2(1-\theta)\sigma i \tan\frac{1}{2}\xi}{1 + 2\theta\sigma i \tan\frac{1}{2}\xi}$ $\qquad\qquad$ (16.28)

It is interesting to see that if ξ is small (say, 30 or more grid points per wave length), which is often the case

$$\sigma\sin\xi \simeq 2\sigma\tan\tfrac{1}{2}\xi \simeq \sigma\xi = 2\pi\Delta t/T$$

which means that the time step is the major factor influencing the accuracy. Moreover, in this approximation the two implicit methods become identical. For this case, Fig. 16.8 gives the damping rates and relative wave speeds for the three methods. Note that for leap-frog the damping rate is exactly unity. The same is true for the implicit methods if the parameter $\theta = \frac{1}{2}$.

The wave speed for the leap-frog method is a little too high. For the implicit methods, the error in the wave speed is of the same order but of opposite sign. The factor θ turns out to have a very great effect in the wave damping; apparently you should use a value equal or close to $\frac{1}{2}$. In practice, 0.52 or 0.55 is often used. The reason is that there is no damping at all for $\theta = \frac{1}{2}$. Any destabilizing effect due to, e.g. nonlinear terms may then lead to instability. Using a slightly higher value of θ gives some numerical damping to take care of such effects.

Which method is now to be recommended for general use? If there are no constraints on the grid size, you may conclude that the leap-frog method is more efficient than the implicit ones: from an accuracy point of view the same time step and grid size can be used in both, and leap-frog requires much less work. Therefore, you would expect the latter to be in general use. That this is not so comes from the fact that you mostly do have constraints on the grid size. If you want to model the

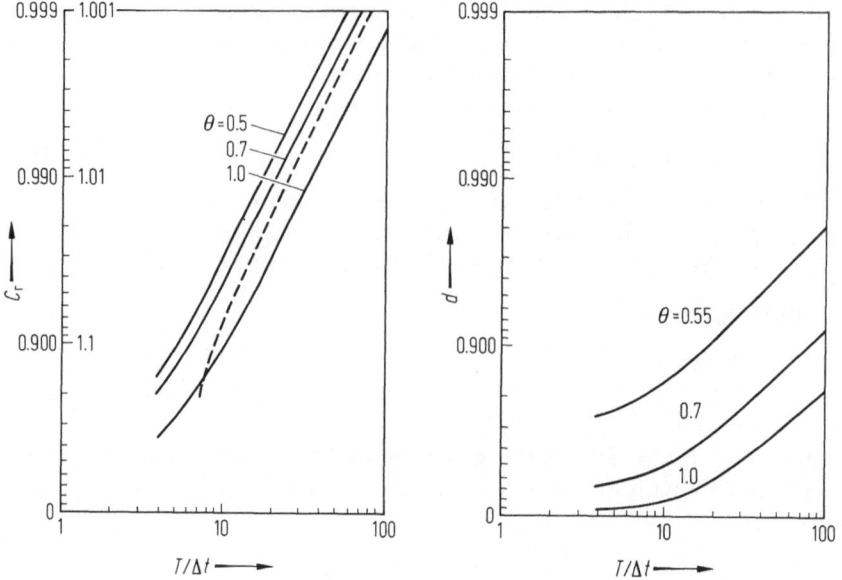

Fig. 16.8 Damping rate d at one wave period, and relative wave speed c_r for leap-frog (dashed), Crank–Nicholson and Preissmann (drawn lines).

geometry of a certain river with a certain detail, a grid size of 1 or 2 km is usually about the maximum. If you realize that the tidal wave length is of the order of several hundred km, the grid size is much smaller than required for wave propagation only. This is still the same for all methods. However, the effect on the time step is different. For leap-frog you have to obey the stability condition (16.10) which leads to a very small time step. The implicit methods do not have this restriction, so the time step can then be chosen purely from wave propagation considerations and be much larger than for leap-frog. Consequently, even though the work per time step is larger, the total amount of work to cover a certain period of time will be smaller. As a result, you will note that most of the standard programs for long waves work with some implicit method.

16.6 Example

In a tidal river, the width and bottom level of which are illustrated in Fig. 16.9, a navigation channel is dredged upto a harbour 60 km from the sea and with an

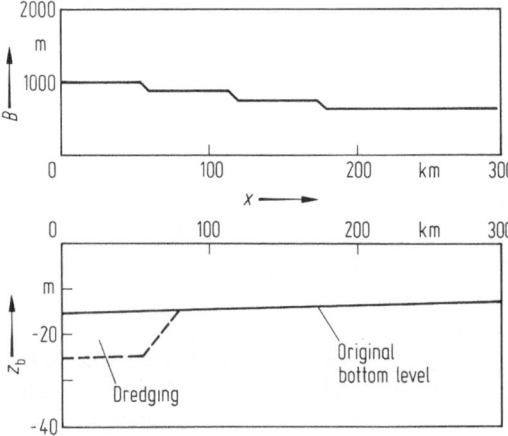

Fig. 16.9 Width and depth of tidal river.

additional depth of 10 m. At the sea, there is a tide with amplitude 1.20 m and period 12.5 h. What is the effect of the dredging on the waterlevels and discharges in the river?

To represent the geometry, a staggered grid with grid size 3 km has been used. This may even be too large for a more realistic case, but it will do for the example. At any rate, it is small enough compared with the tidal wave length of 550 km. To determine the time step, an accuracy criterion has to be stated. For this purpose, it seems reasonable to require an accuracy of 1 cm in the water level at the harbour. The travel time of the tidal wave over a distance of 60 km is about 0.11 T. The number of time steps involved is $n = 0.11$ $T/\Delta t$. You can now try some values of $T/\Delta t$, compute the amplification factor and the damping rate; if the latter is greater than 0.99, the time step is acceptable. The amplitude error upstream of the harbour may then be larger, but on the other hand the tidal wave is damped so the accuracy of 1 cm will be met overall. The phase errors are of no importance in this case. You may check for yourself that the criterion is satisfied for the Crank–Nicholson or Preissmann methods if $T/\Delta t = 20$ or $\Delta t = 0.5$ h. This value has been used in the following results.

Figures 16.10 and 16.11 give the maximum and minimum computed water levels and discharges along the river. As could be expected, the tide penetrates into the river to a greater distance after dredging; a greater volume of water flows in and out accordingly. However, the maximum waterlevel hardly changes as the bottom friction decreases in the dredged region. Therefore, there does not seem to be any danger of flooding for the surrounding area.

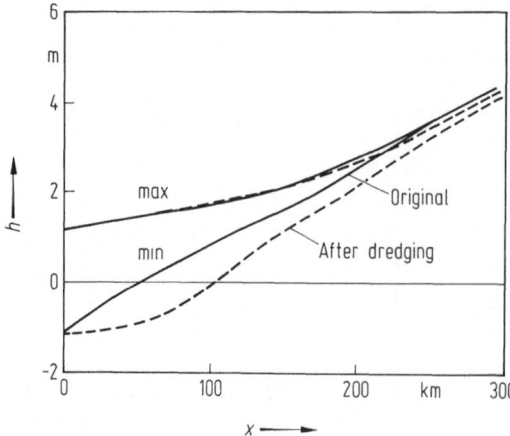

Fig. 16.10 Maximum and minimum waterlevels in the original situation and after dredging.

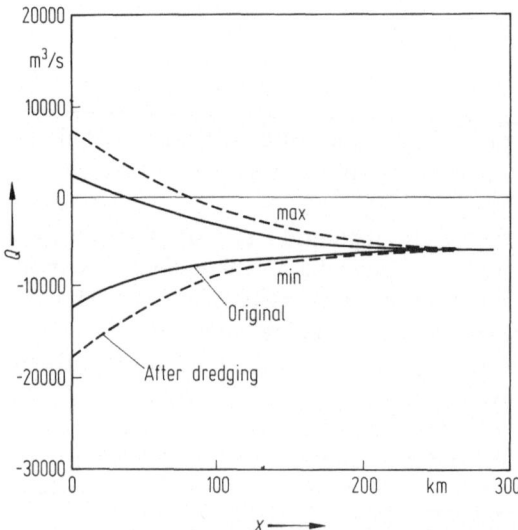

Fig. 16.11 Maximum and minimum discharges in the original situation and after dredging.

16.7 Exercises

1. Show that the leap-frog method is always unstable if you take the friction term in eq. (16.2) at point $(j-1, n+1)$ (that is: show that the amplification factor can exceed unity for some wave length whatever you choose for Δx, Δt).

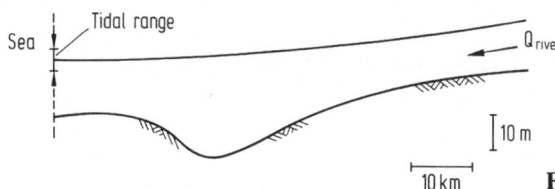

10 km

Fig. 16.12 Example of a tidal river.

2. Investigate the truncation error of the leap-frog method. Which reference point for the Taylor series do you choose in eqs. (16.1) and (16.2)? Develop both unknowns into Taylor series.
3. Determine the stability condition for the leap-frog method with bottom friction. In this case, you will have to compute the roots of eq. (16.7) explicitly and discern between the cases where they are real or complex.
4. A tidal computation is performed for a river with a bottom profile as shown in fig. 16.12. At which point do you evaluate the stability criterion for an explicit method. And at which waterlevel?
5. Verify the amplification factors (16.17) to (16.19) for the Crank–Nicholson and Preissmann methods.
6. Determine the truncation errors for the Crank–Nicholson and Preissmann methods. Which reference points do you take for the Taylor expansions?
7. Verify that the propagation factors for leap-frog according to eqs. (16.8) (with $r = 0$) and (16.27) agree.

Chapter 17
Long Waves in Two-Dimensional Areas

17.1 Mathematical Formulation

The theory of long waves applies also to tidal waves or storm surges in the sea, in shallow coastal areas and estuaries, to flood waves in rivers with flood plains and similar situations. In many such cases, the wave length is so much larger than the water depth that a two-dimensional, depth-averaged mathematical model is adequate. The formulation is essentially the same as in Chapters 15 and 16, if the dependence on two horizontal coordinates x, y is taken into account. See Fig. 17.1 for definitions.

$$\frac{\partial h}{\partial t} + \frac{\partial}{\partial x}(au) + \frac{\partial}{\partial y}(av) = 0 \tag{17.1}$$

$$\frac{\partial u}{\partial t} + u\frac{\partial u}{\partial x} + v\frac{\partial u}{\partial y} - fv + g\frac{\partial h}{\partial x} + ru = 0 \tag{17.2}$$

$$\frac{\partial v}{\partial t} + u\frac{\partial v}{\partial x} + v\frac{\partial v}{\partial y} + fu + g\frac{\partial h}{\partial y} + rv = 0 \tag{17.3}$$

where f is the Coriolis parameter, indicating the effect of the earths' rotation. For simplicity, a linear bottom friction with coefficient r has been introduced. In actual applications, you should of course use the complete equations as given in Appendix A1.2.

It is interesting to note that the shallow-water (or long-wave) equations are very similar to those for a compressible 2-d flow, which read:

$$\frac{\partial \rho}{\partial t} + \frac{\partial}{\partial x}(\rho u) + \frac{\partial}{\partial y}(\rho v) = 0 \tag{17.4}$$

$$\frac{\partial u}{\partial t} + u\frac{\partial u}{\partial x} + v\frac{\partial u}{\partial y} + \frac{1}{\rho}\frac{\partial p}{\partial x} = 0 \tag{17.5}$$

and similarly for the y direction. Here, you need an equation of state to identify the relationship between pressure p and density ρ. If you choose this as $p = \frac{1}{2}g\rho^2$ and

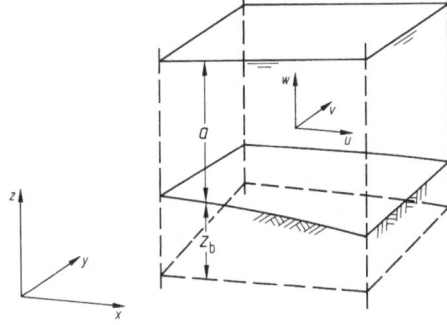

Fig. 17.1 Definitions for the two-dimensional long-wave equations.

identify ρ by a, you get

$$\frac{\partial a}{\partial t} + \frac{\partial}{\partial x}(au) + \frac{\partial}{\partial y}(av) = 0 \tag{17.6}$$

$$\frac{\partial u}{\partial t} + u\frac{\partial u}{\partial x} + v\frac{\partial u}{\partial y} + g\frac{\partial a}{\partial x} = 0 \tag{17.7}$$

which agrees with eqs. (17.1) and (17.2) except for some minor details. In compressible flows, the Mach number is defined as

$$Ma = |v|/c$$

where $c = (\mathrm{d}p/\mathrm{d}\rho)^{1/2}$ indicates the speed of sound waves. In the shallow-water case the corresponding quantity is called the Froude number. With the equation of state just mentioned, $c = (ga)^{1/2}$ as you would expect and as you will see below. The use of this analogy is that you may apply many results (including numerical methods and treatment of boundary conditions, to mention just a few) from the rich experience available for compressible flow in aerodynamics.

17.2 Wave Propagation and Characteristics

It is possible, but very cumbersome, do define characteristics in a two-dimensional sense. They will be planes instead of lines (see the standard text by Courant & Hilbert 1962). In Chapter 15, you have seen that a linearized analysis of wave propagation gives about the same information in a more useful form, so let us apply that in this case. For this purpose, only the essential features from eqs. (17.1) ... (17.3) are taken into account:

– Coriolis and friction are disregarded;
– the convection velocities (i.e. the u, v coefficients in front of the derivatives) are

assumed constant. This amounts to considering disturbances on a uniform base flow as in Chapter 15;
the bottom is assumed to be flat.

Neither of these assumptions is necessary, but they are made to keep the analysis somewhat transparent. If you finally decide to use a reference frame moving with this constant base flow, you get (in vector form)

$$\frac{\partial \mathbf{v}}{\partial t} + A\frac{\partial \mathbf{v}}{\partial x} + B\frac{\partial \mathbf{v}}{\partial y} = 0 \tag{17.8}$$

where

$$\mathbf{v} = \begin{pmatrix} u \\ v \\ a \end{pmatrix} \qquad A = \begin{pmatrix} 0 & 0 & g \\ 0 & 0 & 0 \\ a & 0 & 0 \end{pmatrix} \qquad B = \begin{pmatrix} 0 & 0 & 0 \\ 0 & 0 & g \\ 0 & a & 0 \end{pmatrix}$$

As usual, you can assume wave-like solutions of the form

$$\mathbf{v}(x, y, t) = \mathbf{V} \exp{(ik_1\, x + ik_2 y - i\omega t)} \tag{17.9}$$

where

$\mathbf{V}$ = amplitude vector (complex)

$k_{1,2}$ = wave numbers in x, y directions

ω = angular frequency

The wave length L can be found from

$$L = 2\pi/k \qquad \text{with} \qquad k^2 = k_1^2 + k_2^2$$

The direction of propagation ϕ of the wave is defined by (see Fig. 17.2)

$$k_1 = k \cos\phi$$
$$k_2 = k \sin\phi$$

If you substitute eq. (17.9) into eq. (17.8), you get a homogeneous system of algebraic equations

$$(-i\omega I + ik_1 A + ik_2 B)\mathbf{V} = 0 \tag{17.10}$$

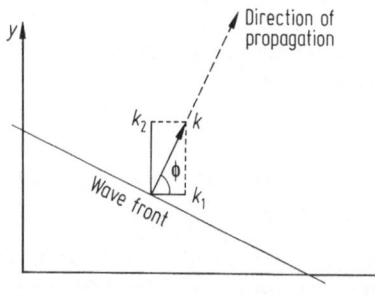

Fig. 17.2 Wave number and direction of propagation.

Given the wave number (vector), this is an eigenvalue problem for ω; you may figure out that this gives

$$\omega(\omega^2 - gak^2) = 0$$

with roots

$$\omega_1 = 0$$
$$\omega_{2,3} = \pm k(ga)^{1/2}$$

(17.11)

For each of these, you can find the eigenvector from the system of equations (17.10) by disregarding one of the three equations. You may check that the eigenvectors are:

$$\mathbf{v}_1 = \begin{pmatrix} \sin\phi \\ -\cos\phi \\ 0 \end{pmatrix} \qquad \mathbf{v}_{2,3} = \begin{pmatrix} \pm\cos\phi \\ \pm\sin\phi \\ (a/g)^{1/2} \end{pmatrix}$$

(17.12)

To understand what this means, consider e.g. the wave corresponding to ω_2:

$$\mathbf{v}(x, y, t) = v_0 \begin{pmatrix} \cos\phi \\ \sin\phi \\ (a/g)^{1/2} \end{pmatrix} \exp(ik_1 x + ik_2 y - i\omega t)$$

This is a wave propagating with a velocity

$$c_2 = \omega_2/k = (ga)^{1/2}$$

in the direction of the wave number vector k (Fig. 17.2). In a fixed reference frame, this is superimposed on the base flow (u, v). Figure 17.3 shows three wave fronts with different directions ϕ at time $t = 0$ and at time t_1. Their displacement normal

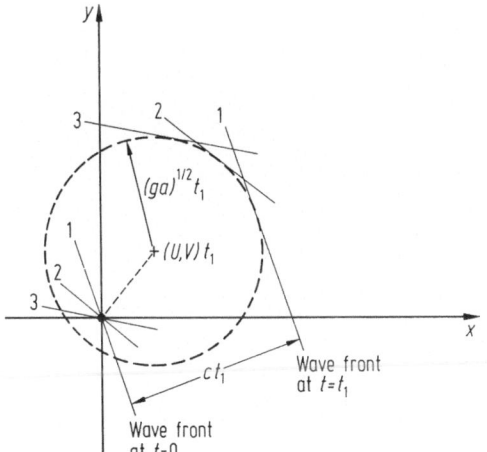

Wave front at $t=t_1$

Wave front at $t=0$

Fig. 17.3 Two-dimensional wave propagation.

to the wave front is

$$ct_1 = \{u_n + (ga)^{1/2}\}t_1 \tag{17.13}$$

where u_n is the component of (u, v) normal to the wave front. You conclude that all these wave fronts are tangent to a circle with radius $(ga)^{1/2}t_1$, of which the centre moves with the base flow velocity.

The variation of velocity in the wave can be chosen arbitrarily at v_0, but its direction is normal to the wave front according to the amplitude vector. The corresponding waterlevel variation is seen to be $v_0 c/g$ and it is in phase with the velocity variation. This agrees completely with the one-dimensional theory (Chapter 15).

The wave corresponding to ω_3 behaves the same way, but moves in the opposite direction relative to the base flow. Its wave fronts are tangent to the same circle as shown in Fig. 17.3. The fact that you find three velocities of propagation, in agreement with the order of the system of differential equations, means that the system is hyperbolic.

The wave with $\omega_1 = 0$ is a special one; it does not exist in one-dimensional theory. The eigenvector in eq. (17.12) indicates that the waterlevel variation is zero for this type of wave, so it is essentially a fixed-surface phenomenon and is observable only in the velocity pattern. The velocity variation in the eigenvector is normal to the wave-number vector, so parallel to the wave front. The speed of propagation is zero, or (u, v) in a fixed frame. Apparently you have a transversal wave moving with the flow velocity (Fig. 17.4). From eq. (17.8), you can check that this wave satisfies

$$\frac{\partial u}{\partial x} + \frac{\partial v}{\partial y} = 0 \tag{17.14}$$

in agreement with the fixed water surface. The quantity transported in this type of waves is vorticity

$$\Omega = \frac{\partial u}{\partial y} - \frac{\partial v}{\partial x} \tag{17.15}$$

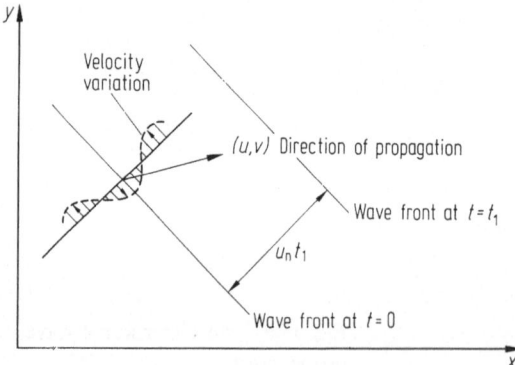

Fig. 17.4 Transversal (vorticity) waves.

which is a measure of the rate of rotation of a fluid particle about its (vertical) axis. If you take the derivatives of eqs. (17.2) and (17.3) with respect to y and x resp. and subtract, you get the equation for the transport of vorticity:

$$\frac{\partial \Omega}{\partial t} + u\frac{\partial \Omega}{\partial x} + v\frac{\partial \Omega}{\partial y} = 0 \qquad (17.16)$$

The pressure, or water level, does not occur in it. The equation confirms that vorticity is transported at the mean flow velocity as you just found. As vorticity cannot be defined in a one-dimensional flow, it will now be obvious, why you do not find the third characteristic in a one-dimensional theory.

17.3 Boundary Conditions

In Chapter 15, you have seen a general rule for hyperbolic problems, stating that the number of boundary conditions, needed at any particular point of the boundary, equals the number of characteristics entering the region at that point. This rule can also be applied in the case of two-dimensional long waves. The characteristics are considered in a direction normal to the boundary. Then, depending on the magnitude and direction of the flow, you can have four different situations:

$u_n \leqslant -(ga)^{1/2}$	no boundary conditions	supercritical outflow
$-(ga)^{1/2} < u_n \leqslant 0$	one b.c.	subcritical outflow
$0 < u_n \leqslant (ga)^{1/2}$	two b.c.	subcritical inflow
$u_n > (ga)^{1/2}$	three b.c.	supercritical inflow

Which situation occurs depends on the sign of the normal flow component, or in plain words: whether the flow is in or out (although the base flow was considered a constant in the preceding section, it is of course a variable in reality). The second factor determining the number of boundary conditions is the Froude number $Fr = u_n/(ga)^{1/2}$. If it exceeds unity in absolute value, you have supercritical flow and you need either no or three boundary conditions. This is not the normal case, however. For subcritical flow, the number of conditions is one (for outflow) or two (for inflow). In tidal flow, this means that the number of boundary conditions varies between ebb and flood flow.

For one of the boundary conditions, it is not difficult to imagine what to specify: this could be the water level or the normal velocity component. However, the second boundary condition on inflow is more difficult. This one is related to the "vorticity wave" discussed above. It would, therefore, be natural to specify the vorticity as a second boundary condition. However, this is a very awkward quantity if the boundary conditions have to be taken from field measurements, as is often the case. On the other hand, it is not unusual that the boundary conditions are taken from a greater (and numerically coarser) model; in that case the vorticity

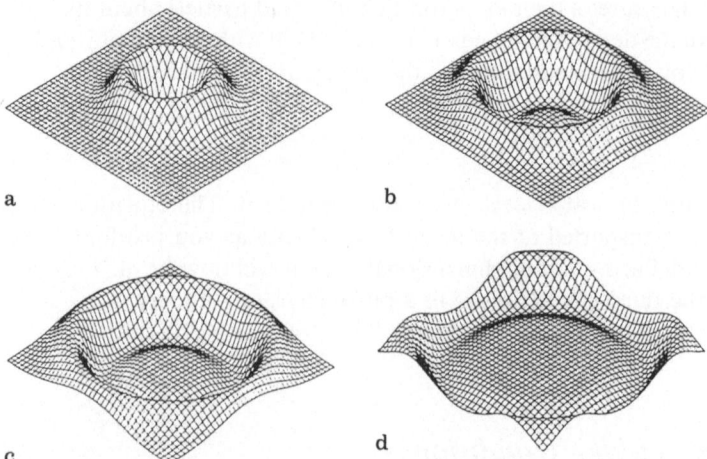

Fig. 17.5 Wave leaving a region with first-order weakly reflecting boundary conditions all around (with permission from Verboom and Slob, 1984).

could be specified. Alternatively, you could specify the tangential velocity component, or the velocity direction as a second condition. All this is not very common; actually, there are many programs for the shallow-water equations which do not require any second boundary condition at all and apparently have some built-in numerical boundary condition. There are indications that the influence of a wrong second boundary condition does not penetrate too far into the region.

As in the one-dimensional case, it is nice to have some sort of non- or weakly reflecting boundary condition, which allows waves to leave the region. This can be done approximately by using exactly the same approach: the 1-d characteristic relation (15.10) is assumed to hold in the direction normal to the boundary. This gives a relation between h and u_n to be used as a boundary condition. As you can expect, it works well for waves passing the boundary approximately in normal direction. At oblique incidence, you get quite some reflection. An example, taken from Verboom and Slob (1984) is shown in Fig. 17.5. There is a systematic technique to derive higher-order non-reflecting boundary conditions (Engquist & Majda 1977), but these get quite complicated. See Verboom and Slob for applications.

17.4 Example

A good idea of the possibilities of two-dimensional long-wave models can be obtained from a relatively old published example by Daubert & Graffe (1967). The model area is the mouth of the Gironde estuary in France (Fig. 17.6). The problem

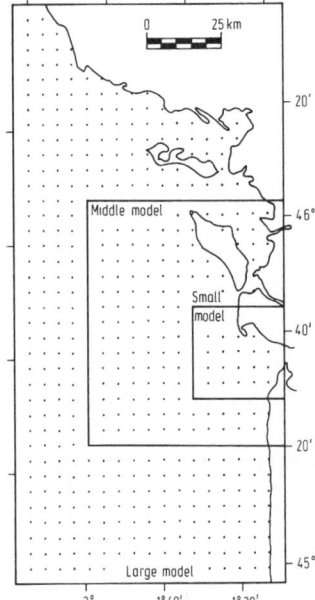

Fig. 17.6 Model boundaries for Gironde model (Daubert & Graffe, 1967).

of getting good boundary conditions was solved here by using three nested models with grid sizes of 5, 2.5 and 1 km respectively. On the boundary of the outer model, a plane incident tidal wave was imposed using the weakly reflecting condition (15.10). The reason of doing this is that the tidal wave will be (partially) reflected; the reflected wave must be allowed to leave the area again. The angle of incidence was determined from field data concerning tidal conditions offshore. At the inward (river mouth) boundary, the water level was imposed as measured in the field. Boundary conditions for the middle model were taken from the computed results of the outer one; in the same way the small model took its conditions from the middle one.

The model was used to do a number of sensitivity experiments (which is easy to do in a numerical model: you can just switch coefficients on or off). It was found for example that switching the convective terms off did not influence the results very much. However, do not consider this as a general conclusion; in situations with more rapid changes in flow direction and/or magnitude they may be quite important. Similarly, a variation in the bottom roughness (i.e. in the friction coefficient c_f) all over the region did not have much influence. This could be expected, since it does no more than vary the rate of damping uniformly at all points. The flow pattern *can* usually be influenced by *local* variations of the friction coefficient. This is one of the common ways of adjusting (or "calibrating") the model and it is justified as the coefficient cannot be measured directly.

Neglecting the Coriolis acceleration (i.e. putting $f = 0$) had quite some effect in the outer model. This is illustrated in Fig. 17.7. The variation of the velocity vector

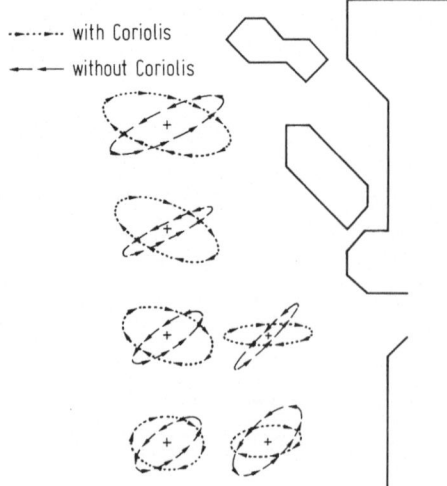

Fig. 17.7 Effect or Coriolis acceleration on tidal flow.

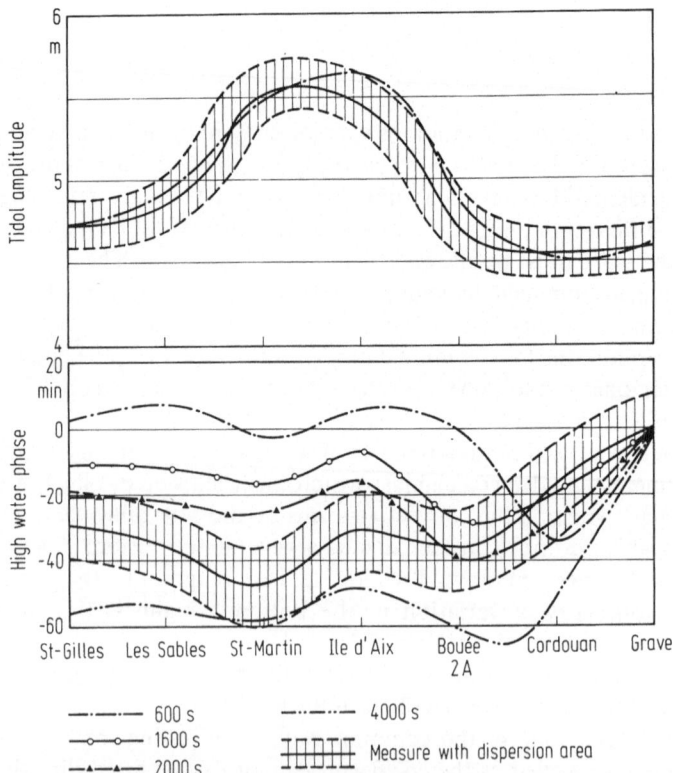

Fig. 17.8 Effect of river boundary phase on tidal levels and phases along the coast.

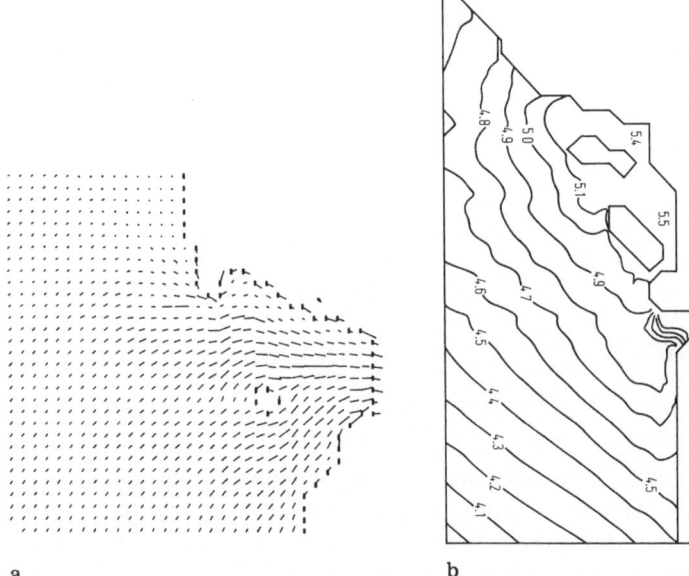

a   b

Fig. 17.9 Computed flow pattern and tidal amplitude (irregularities in contour lines are due to plotting program).

over a tidal cycle is shown there for a few selected points. You can see very clearly that the Coriolis acceleration tends to deviate the flow direction to the right (as it should on the Northern hemisphere). The magnitude of the flow is hardly affected. In the smaller models, neglecting the Coriolis acceleration did not have much influence provided the correct boundary conditions (including Coriolis effects) were put in from the larger models.

Finally, the phase of the river boundary condition relative to the incident tidal wave was adjusted by comparing computed tidal levels and phases with field data at coastal stations (Fig. 17.8). The tidal amplitude was not very much affected, but the phase shift turned out to be rather sensitive. Apart from this, only very few field measurements were available to check the numerical model results. One of the great advantages of the numerical model is that it can produce detailed data at all grid points, as shown, in Fig. 17.9(a, b)

Chapter 18
Finite-Difference Methods for Two-Dimensional Long Waves

18.1 Grids

In order to solve the two-dimensional long-wave equations numerically, the region of interest is covered by a rectangular grid. Recently, curvilinear grids that fit to the boundaries have been used but this is outside the scope of this book. There are various ways of arranging the variables in the grid, three of which are shown in Fig. 18.1.

In grid 1, each grid point has three variables to be computed: the water level h and the two velocity components (u, v). In grid 2, which is staggered, a grid point is either a "waterlevel point" or a "velocity point". If you imagine two such grids and shift one of them over one grid interval up or down, you get grid 1 again. This means that grid 2 economizes on the number of variables by a factor of 2. In grid 3, there is an even further reduction. Again you can take two such grids and shift one over a grid interval in both directions: the result is grid 2. So you have gained another factor 2. Yet all grids have the same grid intervals Δx, Δy. You will not be surprised that grid 3 is used very often.

There is another reason to use grid 3, coming from the structure of the finite-difference equations. If you look at eq. (17.8), to compute a new value of u, you need the waterlevels to the right and left of it and these are exactly the quantities available in grid 3. You may check the other equations similarly.

A final advantage of both grids 2 and 3 is that they cannot support waves with length $2\Delta x$ which may sometimes cause difficulties with stability (see for example section 16.2). The idea is that corresponding (e.g. waterlevel) points are that distance apart and therefore cannot represent waves shorter than $4\Delta x$.

Grid 3 has a disadvantage in representing the Coriolis and convective terms. For the former, you need v in a u-point, which is not there. It can, however, be interpolated between the four surrounding points without loss of accuracy. For the convective terms in point (k, j), you would need u in the neighbouring points $(k + 1, j)$ and $(k - 1, j)$. The only way to get these is interpolating between $(k + 2, j)$, (k, j) and $(k - 2, j)$, but this has the effect that the convective terms are differenced at the double grid size. This is no problem if they are unimportant anyway. But if they are important, you will have to reduce the grid size and you lose the advantage of efficiency of this grid.

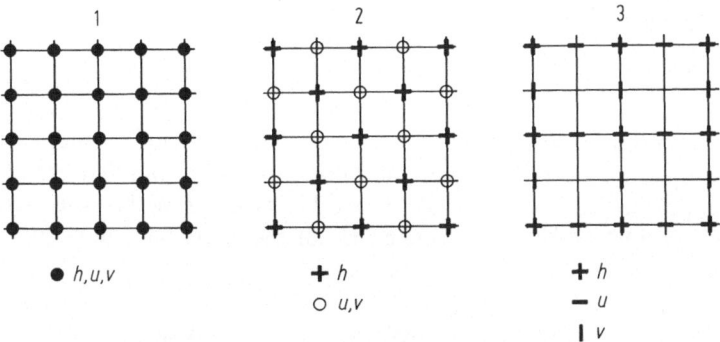

Fig. 18.1 Three possible grid layouts.

Rectangular grids are, of course, not very good in representing curved boundaries. You must use a stepwise linear approximation that fits as well as possible to the real boundary. If a boundary condition for the waterlevel is specified, waterlevel points must be on the boundary. On closed and velocity boundaries, you must have velocity points (and for grid 3 of the correct type, i.e. u or v). Usually, it requires some experimentation to find a good orientation of the grid with respect to all boundaries. Some examples are given in Fig. 18.2.

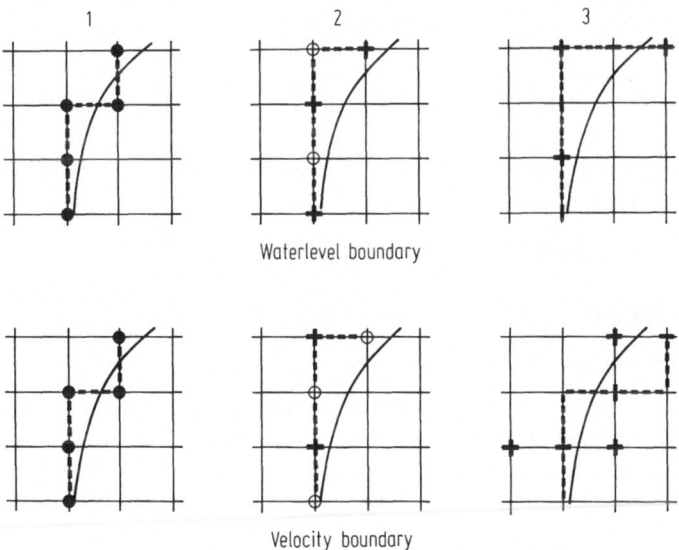

Fig. 18.2 Boundary representation in rectangular grids.

18.2 Explicit Method

There are several possibilities to construct explicit finite-difference methods for the long-wave equations. A relatively popular one is the following that has been used by various people (e.g. Sielecki 1968). For other types of methods see Kinnmark (1986). Using grid 2, and starting from eq. (17.8), the finite-difference equations can be written as

$$u_{k,j}^{n+1} = u_{k,j}^n + \Delta t f v_{k,j}^n - g \frac{\Delta t}{2\Delta x}(h_{k+1,j}^n - h_{k-1,j}^n) - r\Delta t u_{k,j}^n \tag{18.1}$$

$$v_{k,j}^{n+1} = v_{k,j}^n - \Delta t f u_{k,j}^{n+1} - g \frac{\Delta t}{2\Delta y}(h_{k,j+1}^n - h_{k,j-1}^n) - r\Delta t v_{k,j}^n \tag{18.2}$$

$$h_{k,j}^{n+1} = h_{k,j}^n - a\frac{\Delta t}{2\Delta x}(u_{k+1,j}^{n+1} - u_{k-1,j}^{n+1}) - a\frac{\Delta t}{2\Delta y}(v_{k,j+1}^{n+1} - v_{k,j-1}^{n+1}) \tag{18.3}$$

There is a little trick involved in eq. (18.3) by using the new values of the velocities (similarly for u in eq. 18.2). This suggests an implicit method, but it is effectively explicit as these velocity values have already been computed from the other equations. As you will see in section 18.4, this has a beneficial effect on the stability.

18.3 Alternating-Direction Implicit Method

As expected, the explicit method is restricted by stability conditions on the time step (section 18.4). Therefore, implicit methods can be tried, e.g. the Crank–Nicholson method. To keep the notation somewhat tractable, the vector form of eqs. (17.8) is used:

$$\mathbf{v}^{n+1} - \mathbf{v}^n + \theta\Delta t(AD_{ox}\mathbf{v}^{n+1} + BD_{oy}\mathbf{v}^{n+1}) + (1-\theta)\Delta t(AD_{ox}\mathbf{v}^n + BD_{oy}\mathbf{v}^n) = 0 \tag{18.4}$$

where a short-hand notation is used for the differences:

$$D_{ox}\mathbf{v} = (\mathbf{v}_{k+1,j} - \mathbf{v}_{k-1,j})/2\Delta x \tag{18.5}$$

and θ as usual indicates the implicit character. Grid type 3 can conveniently be used in this approach. If you look at this system, you will conclude that it involves as unknowns all variables (h, u, v) at all grid points at the new time level. This may be several thousands of unknowns. Although the system has a band structure and can be solved on present-day computers. it is not a particularly attractive process.

Douglas & Gunn (1964) found an ingenious way out of the problem by splitting the process into two steps. First, auxiliary values are computed with only the x-

derivatives treated implicitly:

$$v^* - v^n + \theta\Delta t(AD_{ox}v^* + BD_{oy}v^n) + (1-\theta)\Delta t(AD_{ox}v^n + BD_{oy}v^n) = 0 \quad (18.6)$$

This is in itself a consistent approximation of the differential equations, but the x and y derivatives are treated differently. The auxiliary values are used to evaluate the x-differences in eq. (18.4), so you get as a final step

$$v^{n+1} - v^n + \theta\Delta t(AD_{ox}v^* + BD_{oy}v^{n+1}) + (1-\theta)\Delta t(AD_{ox}v^n + BD_{oy}v^n) = 0$$
$$(18.7)$$

Note that this looks very much like eq. (18.6) but now the y-derivatives are treated implicitly. The entire process takes implicit steps in x and y alternatively, whence it is called an alternating-direction implicit (ADI) method.

You might wonder what the advantage of all this is. In eq. (18.6), the unknowns are coupled only in lines, with tridiagonal matrices which can be handled efficiently by the Thomas algorithm (section 7.4). Such a (small) system has to be solved for each row separately. In the second step, the columns form tridiagonal systems. So, instead of solving one large system of equations in each time step for eq. (18.4), you now solve a number of small systems in the ADI method.

To consider the truncation error of the ADI method, some more short-hand notation is needed:

$$P = \Delta t\, AD_{ox} = \Delta t\, A\frac{\partial}{\partial x} + O(\Delta t\Delta x^2) \tag{18.8}$$

$$Q = \Delta t\, BD_{oy} = \Delta t\, B\frac{\partial}{\partial y} + O(\Delta t\Delta y^2) \tag{18.9}$$

Using this, eqs. (18.6), (18.7) can be written as

$$(I+\theta P)v^* + \{-I+(1-\theta)P+Q\}v^n = 0 \tag{18.10}$$

$$(I+\theta Q)v^{n+1} + \theta Pv^* + \{-I+(1-\theta)(P+Q)\}v^n = 0 \tag{18.11}$$

Now, if Δt is small, you can use the expansion (which is mathematically not so simple, but hopefully plausible)

$$(I+\theta P)^{-1} \approx I - \theta P + \theta^2 P^2 + \ldots .$$

Then from eq. (18.10)

$$v^* \approx (I-\theta P+\theta^2 P^2)(I-(1-\theta)P-Q)v^n$$
$$\approx (I-P-Q+\theta^2 P^2+\theta PQ+\ldots)v^n$$

and substituting this into eq. (18.11) you get for a complete time step, up to quadratic terms in Δt:

$$v^{n+1} \approx \{I-P-Q+\theta(P^2+PQ+QP+Q^2)+\ldots\}v^n \tag{18.12}$$

On the other hand, the exact solution at time $n+1$ can be developed into a Taylor series; using eq. (17.8) to eliminate the time derivatives, you get

$$\mathbf{v}^{n+1} = \mathbf{v}^n + \Delta t \frac{\partial \mathbf{v}}{\partial t} + \frac{1}{2} \Delta t^2 \frac{\partial^2 \mathbf{v}}{\partial t^2} + \cdots$$

$$= \mathbf{v}^{\dot{n}} - \Delta t \left(A \frac{\partial \mathbf{v}}{\partial x} + B \frac{\partial \mathbf{v}}{\partial y} \right) + \frac{1}{2} \Delta t^2 \left(A^2 \frac{\partial^2 \mathbf{v}}{\partial x^2} + AB \frac{\partial^2 \mathbf{v}}{\partial x \partial y} + \right.$$

$$\left. + BA \frac{\partial^2 \mathbf{v}}{\partial x \partial y} + B^2 \frac{\partial^2 \mathbf{v}}{\partial y^2} \right) + \cdots \tag{18.13}$$

Keeping the definition of P and Q in mind, this agrees with eq. (18.12) if $\theta = \frac{1}{2}$. Then the method is accurate to second order in time. Otherwise, there remains (after division by Δt to conform to the differential form) a truncation error of first order

$$\frac{1}{\Delta t} \left(\theta - \frac{1}{2} \right) (P^2 + PQ + QP + Q^2) \mathbf{v} = \left(\theta - \frac{1}{2} \right) \Delta t g a \begin{pmatrix} \dfrac{\partial^2 u}{\partial x^2} & + & \dfrac{\partial^2 v}{\partial x \partial y} \\[2mm] \dfrac{\partial^2 u}{\partial x \partial y} & + & \dfrac{\partial^2 v}{\partial y^2} \\[2mm] \dfrac{\partial^2 h}{\partial x^2} & + & \dfrac{\partial^2 h}{\partial y^2} \end{pmatrix} \tag{18.14}$$

For an unsteady situation, this introduces a numerical error that has some resemblance to diffusion (particularly in the third equation). The importance is difficult to estimate; see section 18.5, for another approach to accuracy. For steady flow, the entire expression (18.14) vanishes and the remaining errors are those from the spatial (central) differences.

The ADI method can be put into another convenient form by defining a half time-step value

$$\mathbf{v}^{n+\frac{1}{2}} = \frac{1}{2}(\mathbf{v}^* + \mathbf{v}^n) \tag{18.15}$$

(remember that $\mathbf{v}^*$ is an approximation of the full time-step result). Then with $\theta = \frac{1}{2}$, eqs. (18.6) and (18.7) are rewritten as

$$\mathbf{v}^{n+\frac{1}{2}} - \mathbf{v}^n + \frac{1}{2} \Delta t \, A D_{ox} \mathbf{v}^{n+\frac{1}{2}} + \frac{1}{2} \Delta t B D_{oy} \mathbf{v}^n = 0 \tag{18.16}$$

$$\mathbf{v}^{n+1} - \mathbf{v}^{n+\frac{1}{2}} + \frac{1}{2} \Delta t \, A D_{ox} v^{n+\frac{1}{2}} + \frac{1}{2} \Delta t B D_{oy} \mathbf{v}^{n+1} = 0 \tag{18.17}$$

This can be interpreted as two half time steps in which one derivative is treated explicitly and the other implicitly. In each step, only two levels of the vector $\mathbf{v}$ are needed. This requires less computer storage than eqs. (18.6), (18.7) where three levels are involved. In the form of eqs. (18.16), (18.17), the method has been put forward by Leendertse (1967) and widely used.

18.4 Stability

For the analysis of stability, you can use the Von Neumann method again. It is somewhat more complicated in two dimensions than in one, but the priniple is exactly the same. In order not to burden the discussion too much, Coriolis, bottom friction and convection are disregarded; in this way you get necessary, but perhaps not sufficient conditions. For a more detailed analysis see, e.g., Kinnmark (1986). The conditions derived here give at least good indications.

As an initial condition, take a harmonic wave in two dimensions:

$$v^0 = V \exp (ik_1 x + ik_2 y) \tag{18.18}$$

If you substitute this into the finite-difference equations (18.1) . . . (18.3), you get

$$Cv^1 = DV \exp (ik_1 x + ik_2 y)$$

in which the matrices look like (please check)

$$
C = \begin{pmatrix}
1 & 0 & 0 \\
0 & 1 & 0 \\
a\dfrac{\Delta t}{\Delta x} i \sin \xi & a\dfrac{\Delta t}{\Delta y} i \sin \eta & 1
\end{pmatrix}
$$

$$
D = \begin{pmatrix}
1 & 0 & -g\dfrac{\Delta t}{\Delta x} i \sin \xi \\
0 & 1 & -g\dfrac{\Delta t}{\Delta y} i \sin \eta \\
0 & 0 & 1
\end{pmatrix}
$$

and

$$\xi = k_1 \Delta x \qquad \eta = k_2 \Delta y$$

In each time step, the vector v is multiplied by a matrix $G = C^{-1} D$ which is called the *amplification matrix*. After n steps, the numerical solution is

$$v^n = G^n \exp (ik_1 x + ik_2 y) \tag{18.19}$$

The amplification matrix will normally have three eigenvalues ρ_i with corresponding eigenvectors v_i. Then the amplitude vector V can be written as a linear combination of the three eigenvectors:

$$v = \sum_{i=1}^{3} \alpha_i v_i$$

Multiplication by the matrix G gives

$$GV = \sum_{i=1}^{3} \alpha_i G v_i = \sum_{i=1}^{3} \alpha_i \rho_i v_i$$

so in general

$$v^n = \sum_{i=1}^{3} \alpha_i \rho_i^n v_i \exp(ik_1 x + ik_2 y) \tag{18.20}$$

This indicates that the solution may grow without bound if any of the eigenvalues (amplification factors) exceeds unity in absolute value. Therefore, according to Von Neumann, for stability all the eigenvalues must satisfy

$$|\rho_i| \leqslant 1 \qquad \text{for all} \qquad \xi, \eta \tag{18.21}$$

This is a straightforward extension of the conditions you have seen before. To compute the eigenvalues you do not have to compute the inverse matrix C^{-1} as it follows from the definition that

$$C^{-1} D v_i = \rho_i v_i \qquad \text{or} \qquad (D - \rho_i C) v_i = 0 \tag{18.22}$$

The determinant of this homogeneous system must be zero:

$$\begin{vmatrix} 1-\rho & 0 & -g\dfrac{\Delta t}{\Delta x} i \sin \xi \\[2ex] 0 & 1-\rho & -g\dfrac{\Delta t}{\Delta y} i \sin \eta \\[2ex] \rho a\dfrac{\Delta t}{\Delta x} i \sin \xi & \rho a\dfrac{\Delta t}{\Delta y} i \sin \eta & 1-\rho \end{vmatrix} = 0$$

which can be elaborated to

$$(\rho - 1)\{\rho^2 - (2 - \sigma_x^2 \sin^2 \xi - \sigma_y^2 \sin^2 \eta)\rho + 1\} = 0 \tag{18.23}$$

in which you have now two Courant numbers $\sigma_x = (ga)^{1/2} \Delta t / \Delta x$ and $\sigma_y = (ga)^{1/2} \Delta t / \Delta y$. There is one root $\rho = 1$ which satisfies the condition of stability. The other two may be real or conjugate complex, with their product equal to 1. If they are real, one of them will exceed unity and the method is unstable. So for stability, the roots must be complex, which is the case if

$$(2 - \sigma_x^2 \sin^2 \xi - \sigma_y^2 \sin^2 \eta)^2 \leqslant 4$$

or

$$\sigma_x^2 \sin^2 \xi + \sigma_y^2 \sin^2 \eta \leqslant 4$$

This must still be true for the most unfavourable values of ξ, η which are obtained for $\sin \xi = \sin \eta = 1$. Then you find the stability condition for the explicit method:

$$\sigma_x^2 + \sigma_y^2 \leqslant 4 \tag{18.24}$$

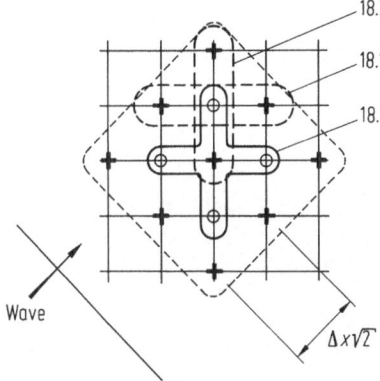

Fig. 18.3 Numerical dependence over full time step for explicit method.

If $\Delta x = \Delta y$, this gives $\sigma^2 \leqslant 2$ which seems to deviate from the CFL condition in the one-dimensional case ($\sigma^2 \leqslant 1$). This is explained in Fig. 18.3 where it is indicated how h^{n+1} is obtained from h^n. Due to the implicit character of the continuity equation, the intermediate velocity values are of no interest here. Imagine a wave propagating at 45° across the grid. It meets grid points (for water level) at distances $2^{1/2}\Delta x$; the CFL condition then becomes

$$2^{-1/2}(ga)^{1/2}\Delta t/\Delta x \leqslant 1$$

which is exactly what you just found. You may convince yourself that this is the most unfavourable wave direction. If the equation of continuity would not have been treated implicitly, you would have to take the distance between (u, v) and h points into account, which is smaller by a factor of 2 and therefore would also give a smaller time step from the CFL condition.

For the ADI method, the reasoning is the same. For each eigenvector $\mathbf{v}_i$, it is convenient to introduce an auxiliary vector $\mathbf{v}^*$; the vector $\mathbf{v}^{n+1}$ can then be replaced by $\rho_i \mathbf{v}_i$. The amplification matrix is a different one from the explicit method, but again it is not necessary to compute it. To determine the eigenvalues, you just combine all 6 equations (18.6), (18.7) with the substitutions just mentioned:

$$\begin{pmatrix} 1 & 0 & -(1\text{-}\theta)pg & -1 & 0 & -\theta pg \\ 0 & 1 & -qg & 0 & -1 & 0 \\ -(1-\theta)pa & -qa & 1 & -\theta pa & 0 & -1 \\ \rho-1 & 0 & (1-\theta)pg & 0 & 0 & \theta pg \\ 0 & \rho-1 & (\theta\rho+1-\theta)qg & 0 & 0 & 0 \\ (1-\theta)pa & (\theta\rho+1-\theta)qa & \rho-1 & \theta pa & 0 & 0 \end{pmatrix} \begin{pmatrix} \mathbf{v}_i \\ \mathbf{v}_i^* \end{pmatrix} = 0$$

$$(18.25)$$

where $p = \Delta t/\Delta x\, i\sin\xi$ and $q = \Delta t/\Delta y\, i\sin\eta$.

Again, the determinant of this system must be zero, which gives with considerable algebra

$$\rho_1 = 1$$

and $\rho_{2,3}$ from

$$\left(\frac{\rho-1}{\theta\rho+1-\theta}\right)^2 = -\sigma_x^2 \sin^2\xi - \sigma_y^2 \sin^2\eta - \theta^2\sigma_x^2\sigma_y^2 \sin^2\xi \sin^2\eta \qquad (18.26)$$

Write the right-hand side as z^2 where $z = z_1 + iz_2$. Then

$$\rho = \frac{1+(1-\theta)z}{1-\theta z}$$

and

$$|\rho|^2 = \frac{\{1+(1-\theta)z_1\}^2 + (1-\theta)^2 z_2^2}{(1-\theta z_1)^2 + \theta^2 z_2^2}$$

This must be less than unity for stability, which gives

$$2z_1 + (1-2\theta)|z|^2 \leqslant 1$$

This is certainly satisfied if $z_1 = 0$ and $\theta \geqslant \frac{1}{2}$. The former is true, as you can see from eq. (18.26), so a sufficient condition for stability (as in the one-dimensional case) is $\theta \geqslant \frac{1}{2}$, regardless of the value of the time step. If $\theta = \frac{1}{2}$ (which is the case for the Leendertse type methods), you can figure out that $|\rho| = 1$, which means that stability is only marginal: there is no wave amplification, but no attenuation either. In such cases, the (physical) bottom friction must ensure stability.

18.5 Wave Propagation

From the Von Neumann stability analysis, it is only a small step to numerical wave propagation. Referring to section 17.2, the exact solution for a wave with wave number k and direction ϕ is

$$\mathbf{v}(x, y, t) = \sum_{j=1}^{3} \alpha_j \mathbf{V}_j \exp(ik_1 x + ik_2 y - i\omega_j t) \qquad (18.27)$$

where $\mathbf{V}_j$ are the eigenvectors given in eq. (17.12). The coefficients α_j indicate the composition of the wave from the three basic wave types.

The corresponding numerical solution was found in the previous section:

$$\mathbf{v}(x, y, t_n) = \sum_{j=1}^{3} \beta_j \mathbf{v}_j \rho_j^n \exp(ik_1 x + ik_2 y) \qquad (18.28)$$

with eigenvectors v_j of the amplification matrix. There are three possible differences between the exact and numerical solutions:

(i) The eigenvectors v_j may differ from the exact ones V_j, which means that there may be a different relation between the amplitudes of water level and velocity.
(ii) The coefficients α and β differ (as a consequence of (i)): the three wave types are present in a different "mix".
(iii) The factor ρ'' may differ from $\exp(-i\omega t)$ in amplitude and phase. This is the same type of numerical error that you have seen several times before in this book.

The first type of error is characterized by the following dimensionless numbers:

(a) amplitude ratio $H/V = (g/a)^{1/2} h/(u^2 + v^2)$ \hfill (18.29)

between the water-level and velocity amplitudes, where (h, u, v) are now the components of the eigenvectors v_j. Exact values from eq. (17.12) are 0 and ± 1 respectively.
(b) direction $V/U = v/u$. Exact values:- $\cotan \phi$ and $\tan \phi$
(c) phase difference between h and u is ϕ_{hu}. Exact values from eq. (17.12) are indefinite and 0 respectively.
(d) phase difference between u and v is ϕ_{uv}. Exact values 0 for all waves.

The second type of error is not so easy to quantify; you will have to compute it for a particular case.

The third type of error is characterized by the usual damping factor d and relative wave speed c_r (e.g. section 16.5). These are illustrated, together with the four quantities defined above, in Figs. 18.4 through 18.8. As the vorticity transport is not described very well by these linearized equations, it is not included. The figures therefore refer to the "normal" type of gravity waves occurring in one dimension as well. It turns out that the grid size (if not too small with respect to the relevant wave length, say more than 20 points per wave length) does not have very much influence. Therefore, only the effect of the time step is shown.

For the explicit method, the eigenvectors can be determined from eq. (18.22) with the result

$$p_{2,3} = \gamma \pm i(1 - \gamma^2)^{1/2} \qquad v_{2,3} = \begin{pmatrix} i\sigma_x \sin \xi \\ i\sigma_y \sin \eta \\ (1-p)(a/g)^{1/2} \end{pmatrix} \qquad (18.30)$$

with $\gamma = 1 - \tfrac{1}{2}\sigma_x^2 \sin^2 \xi - \tfrac{1}{2}\sigma_y^2 \sin^2 \eta$

The same procedure can be followed for the ADI method. However, the expressions get so complicated that they are not given here; the result is given in the illustrations.

Some remarks can be made with respect to these results for wave propagation.

(i) The explicit method does not produce any wave damping; neither does the ADI method if $\theta = \tfrac{1}{2}$. For greater values of θ, a quite significant numerical damping is observed. This agrees with the one-dimensional case.

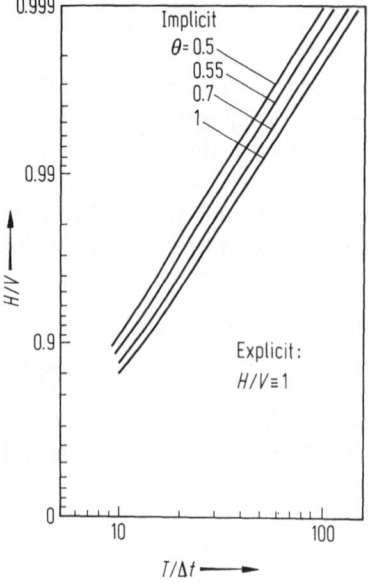

Fig. 18.4 Water-level to velocity amplitude ratio.

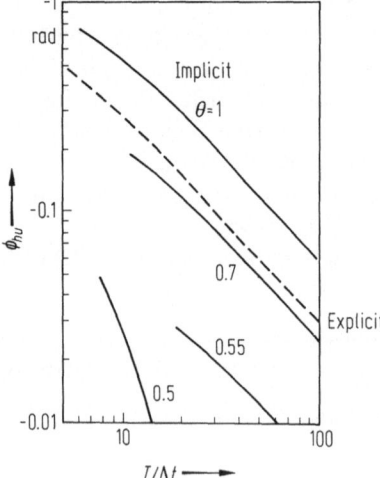

Fig. 18.5 Phase difference between water level and velocity.

(ii) Except for the wave damping and the H/V ratio, the accuracy of the explicit and implicit methods is not too different for comparable time steps; however, you should realize that large time steps are often precluded by stability considerations in the explicit case.

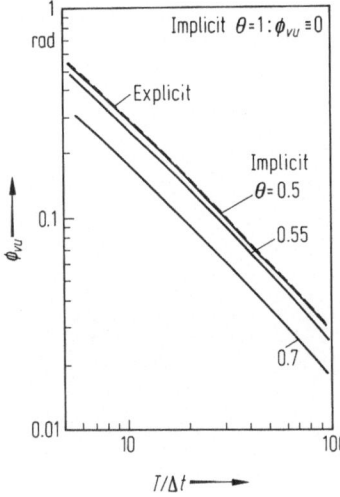

Fig. 18.6 Phase difference between velocity components.

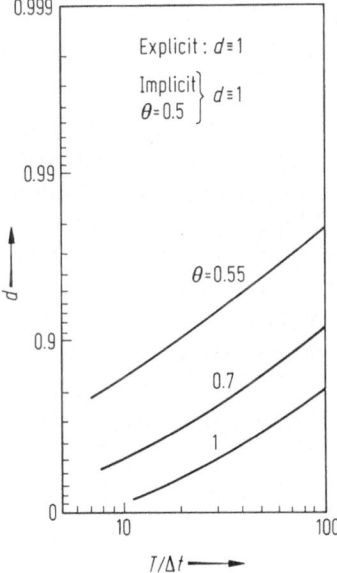

Fig. 18.7 Wave damping per wave period.

(iii) The factor θ has a very great influence on the damping rate. You should be very careful in using values exceeding, say, 0.55. The influence on the other quantities is modest.

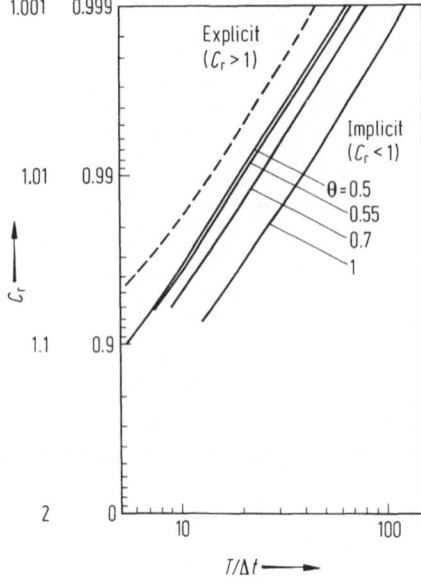

$T/\Delta t \longrightarrow$

Fig. 18.8 Relative wave speed.

(iv) Although not shown in the figures, it turns out that the direction of propagation ϕ has only a limited influence. It is not useful to take it into account in this type of (necessarily rather crude) estimates of accuracy.

Conclusion (ii) could indicate that the explicit method is preferable, as it requires less work per time step. This is correct if wave propagation is the only criterion. However, if one has to take smaller grid sizes in order to represent the bottom and boundary geometry in sufficient detail, this would lead to proportionally smaller time steps for the explicit method, due to stability conditions. For the ADI method, a larger time step may then still be acceptable for accuracy and for stability. As this is the most common situation, many standard computer programs are based on an implicit method of the ADI type. Compare the discussion in section 16.5.

18.6 Example

In a shallow lake, shown schematically in Fig. 18.9, it is considered to construct a dam and bridge across the narrow part. Due to storms, considerable currents may occur in the lake. If the cross-section is constricted by a dam, these currents will increase. A numerical model can be used to investigate this quantitatively. How should such a model be set up?

The depth of the lake is assumed to be 12 m uniformly. The bottom friction coefficient $c_f = 0.003$. The most serious storms are supposed to have a north-

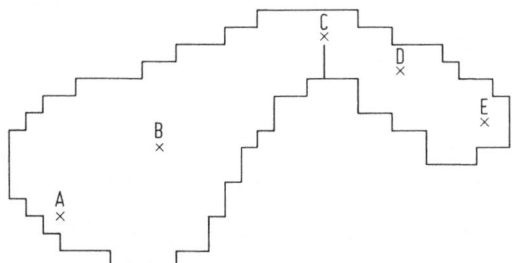

Fig. 18.9 Layout of lake with proposed dam and measuring points.

easterly direction and a wind speed of about $W = 30$ m/s. The wind exerts a shear stress on the water surface of

$$\tau = 0.0035\ W^2$$

in the wind direction. This is added to the shallow-water equations as an external force. The lake is at rest prior to time $t = 0$, when the wind is assumed to start impulsively and to remain constant subsequently. It will be evident to you that many of these assumptions are oversimplifications, but they are sufficient for the purpose of illustration and can be replaced by more realistic data in applications.

What will happen in the lake at a sudden storm is that a transient wave is generated. It develops into a standing wave which will be damped. If the wind would remain constant for a sufficiently long time, a steady state would be reached, but this will usually not be the case. The most serious currents will probably occur soon after the start of the storm. The standing wave is such that the flow is zero at the extremes; the waterlevel variation will be at a maximum there. The basic wave length is therefore twice the basin length, i.e. a little over 60 km. With the given depth, you can compute the wave period as about 1.6 h.

A grid size of 1 km is sufficient for a first, crude representation of the lake geometry. The outline of the lake with this grid size is the one shown in Fig. 18.9. Suppose you have a computer program available that uses the ADI method discussed in the preceding sections, then it remains to select a time step. The factor θ is taken as 0.55 to avoid stability problems. The choice of the time step is based on the accuracy with which the standing wave and its damping rate are reproduced. The propagation speed is quite important in this case, as it determines the wave period from the fixed wave length. You can select some values for the time step and read the corresponding errors from Figs. 18.4 to 18.8, for example:

Δt	1200 s	200 s
$T/\Delta t$	5	27
H/V	0.6	0.98
ϕ_{hu}	0.6	0.02
ϕ_{vu}	0.5	0.1
d	0.7	0.93
c_r	0.85	0.995

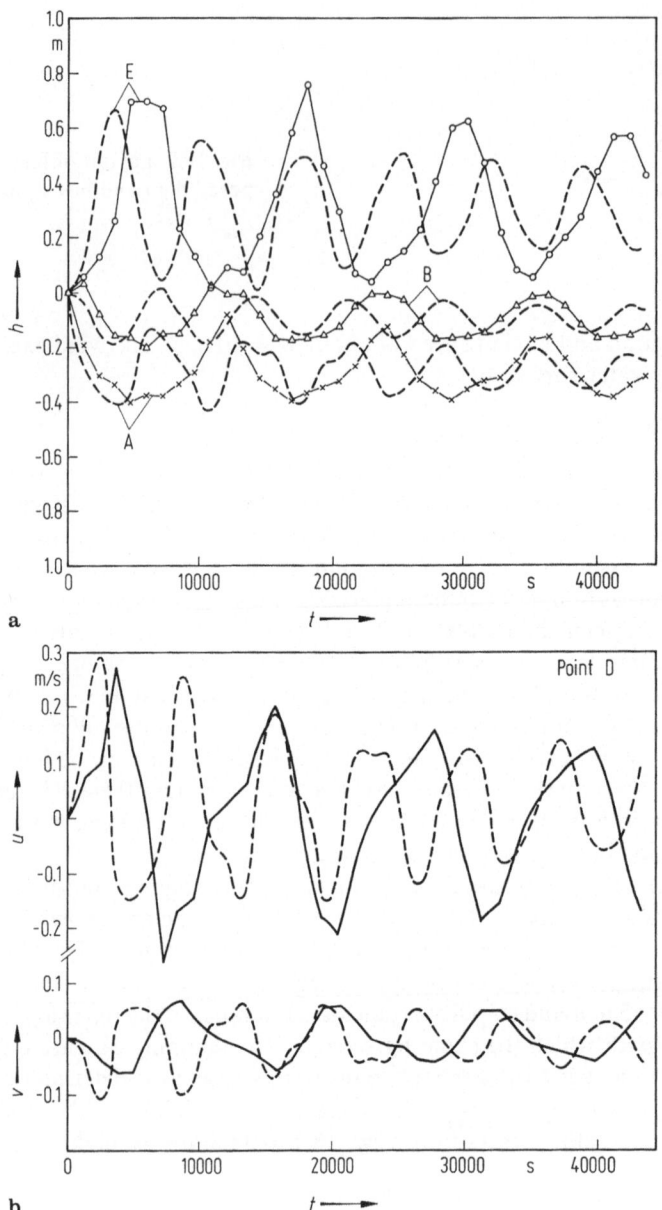

Fig. 18.10 Computed time histories for waterlevel and u (west–east) and v (south–north) velocities for time step 1200 s (drawn) and 200 s (dashed lines).

The values for the first case are tentative only, as the figures are not strictly valid for such small numbers of time steps per wave period. Anyway, it is evident that there is an error of at least 10% in the wave speed, so a comparable error in the wave period can be expected. Moreover, a quite serious numerical damping is found. The other parameters are inaccurate as well. For the second case, the wave speed is quite accurate. There is still some numerical damping. The other quantities have acceptable values.

The two cases are illustrated in Fig. 18.10, where time histories of water level and velocity are shown for some of the points indicated in Fig. 18.9. In the water levels, it is clear that the wave period varies from about 3 hours for the coarse time step to about 1.7 for the fine one. The latter value is in reasonable agreement with the estimate given above. Apparently, the phase error for the coarse time step is even greater than estimated. The lines are too irregular to check the damping rate with any confidence. The same behaviour is found in the velocity records. The amplitude of the waterlevel and velocity variations is also too large in the coarse time step case. In general, it is confirmed that the time step of 1200 s is much too large for a reliable computation.

For the accurate computation, Fig. 18.11 (page 136) shows the flow pattern at time 2400 s, when the currents are about at their maximum. Flow velocities up to about 0.5 m/s are found. With the same numerical data, the situation with the dam has been computed. The flow pattern at time 2400 s is included in Fig. 18.11. The flow velocities have increased to about 0.65 m/s which is a modest but noticeable change. You should realize, however, that this is a very crude schematization. It is possible that local changes near the dam are greater. A computation with a more detailed grid will give more information about the local flow pattern.

18.7 Exercises

1. Show that the explicit method of eqs. (18.1) . . . (18.3) is of second order accuracy in space and first order in time.
2. Analyse what happens to the stability criterion of the explicit method if you do not use values at time level $n + 1$ at the right-hand sides of eqs. (18.2) and (18.3).
3. In the example of section 17.4, grid sizes of 5, 2.5 and 1 km were used. What is the consequence for the time steps if you use an explicit method? Suppose that the average depth is something like 20 m. If you would use the ADI method instead, how would you choose the time step? Would it be influenced by the grid size?

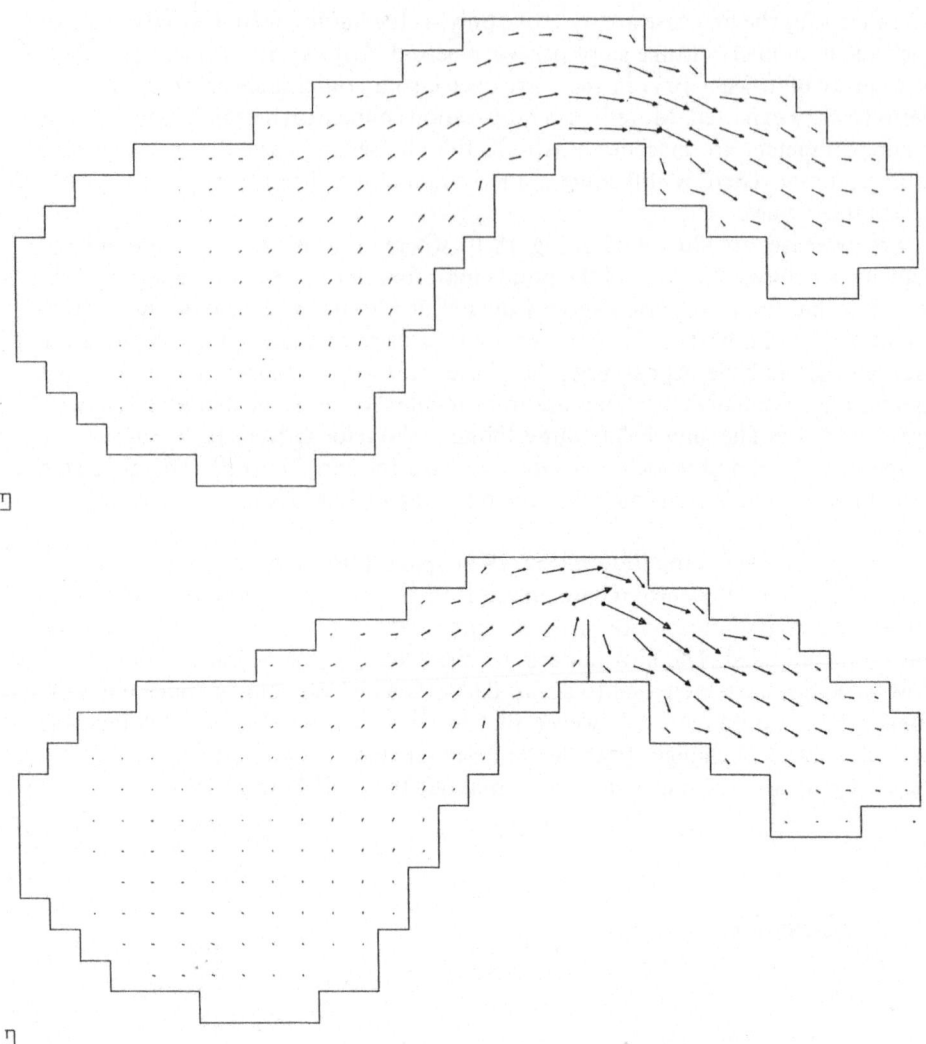

Fig. 18.11 Computed flow patterns at time 2400 s for case without dam (top) and with dam (bottom).

Chapter 19
Potential Flow

19.1 Irrotational Flow

The flow of an incompressible fluid is generally described mathematically by the Navier–Stokes equations (14.9) and (14.11) (in two dimensions; otherwise there is an additional, similar equation), together with the equation of continuity

$$\frac{\partial u}{\partial x} + \frac{\partial v}{\partial y} = 0 \tag{19.1}$$

The numerical solution of these equations is not discussed in this book. There are, however, certain situations in which they can be considerably simplified. Consider a quantity called the vorticity $\omega = \partial u/\partial y - \partial v/\partial x$, which is a measure of the rotation of a fluid particle about its axis (normal to the plane of flow; in 3-d the vorticity is a vector). If the fluid is essentially frictionless and if no vorticity is introduced at the boundaries, it can be shown from the Navier–Stokes equations that

$$\omega = \frac{\partial u}{\partial y} - \frac{\partial v}{\partial x} = 0 \tag{19.2}$$

The flow is then called irrotational and you may be surprised that it is then described by the two simple, linear, equations (19.1) and (19.2), instead of the complicated and nonlinear Navier–Stokes equations.

An example of approximately irrotational flow is the flow about a streamlined object such as an air plane or a ship. As it is the object which moves and not the fluid, there is no inflow of vorticity. Near the object, in a boundary layer, there certainly is a production of vorticity, but outside this layer, the flow is irrotational to a good approximation.

Another interesting example is that of small-amplitude water waves. In this case, it can be shown from an order-of-magnitude analysis that frictional effects in the Navier–Stokes equations are small compared with the local acceleration terms (the first in eqs. 14.9 and 14.11). This means that you can consider the fluid as approximately frictionless and therefore irrotational.

19.2 Potential and Stream Function

If the vorticity is zero, you can imagine a *potential function* $\phi(x, y)$ defined by

$$u = -\frac{\partial \phi}{\partial x} \qquad v = -\frac{\partial \phi}{\partial y} \tag{19.3}$$

You may check that this satisfies eq. (19.2) whatever ϕ is (provided it can be differentiated twice). So instead of finding two unknowns (u, v), you can satisfy yourself by finding only the potential. This is done by substituting eq. (19.3) into eq. (19.1):

$$\frac{\partial^2 \phi}{\partial x^2} + \frac{\partial^2 \phi}{\partial y^2} = 0 \tag{19.4}$$

which is known as the potential or Laplace's equation.

The other way round, you can define a *stream function* ψ defined by

$$u = \frac{\partial \psi}{\partial y} \qquad v = -\frac{\partial \psi}{\partial x} \tag{19.5}$$

(watch carefully the difference with eq. 19.3). This satisfies eq. (19.1) exactly, even if the vorticity is not zero. However, for irrotational flow, you find from eq. (19.2)

$$\frac{\partial^2 \psi}{\partial x^2} + \frac{\partial^2 \psi}{\partial y^2} = 0 \tag{19.6}$$

The stream function turns out to satisfy the same equation as the potential. Yet, the two are really different as they have different boundary conditions. For a much more extensive discussion of irrotational flow see, e.g., the classical text by Lamb (1963).

If you have an arbitrary function $f(x, y)$, a line of constant f is defined by

$$f(x, y) = \text{const.}$$

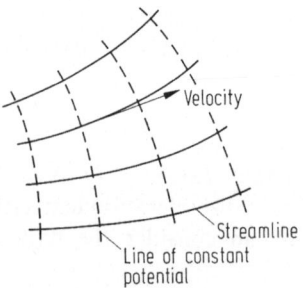

Velocity

Streamline

Line of constant potential

Fig. 19.1 Streamlines and lines of constant potential.

This is usually a curve in the (x, y) plane, the tangent to which is defined by

$$\frac{\partial f}{\partial x}(x - x_0) + \frac{\partial f}{\partial y}(y - y_0) + f(x_0, y_0) = 0$$

where $(\partial f/\partial x, \partial f/\partial y)$ is the normal vector to the curve. Using this and keeping eq. (19.5) in mind, you see that the normal to a line of constant ψ is also normal to the flow direction; in other words: a line of constant ψ is tangent to the flow direction: it is a streamline (hence the name stream function). Similarly, using eq. (19.3), you can conclude that lines of constant ϕ are normal to the flow direction (and consequently to the streamlines), see Fig. 19.1.

19.3 Characteristics and Boundary Conditions

The potential equation is a kind of equilibrium equation: time does not play a part in it, otherwise than (possibly) as a parameter. If you are looking for characteristics, they must be in the (x, y) plane. Following the procedure of section 15.2, you multiply eqs. (19.1) and (19.2) with unknown factors m and n:

$$m\frac{\partial u}{\partial x} + n\frac{\partial u}{\partial y} - n\frac{\partial v}{\partial x} + m\frac{\partial v}{\partial y} = 0 \tag{19.7}$$

and try to get it in the form

$$m\left(\frac{\partial u}{\partial x} + c\frac{\partial u}{\partial y}\right) - n\left(\frac{\partial v}{\partial x} + c\frac{\partial v}{\partial y}\right) = 0 \tag{19.8}$$

where c is now the slope of the lines in the (x, y) plane (so dimensionless). Comparing eqs. (19.7) and (19.8), you find that

$$n/m = - m/n = c$$

and this is not possible for real m and n. The conclusion is that characteristics do not exist for the potential equation. This type of equation is called *elliptic*.

You can get some idea of the type of solutions from a few special cases. For example, the potential of a point source of strength c is

$$\phi(x, y) = c \ln(r/r_0) \tag{19.9}$$

where r is the distance between the location of the source (x_0, y_0) and the location (x, y) where you compute the potential. The streamlines are straight lines, directed radially from the source (Fig. 19.2). If you compute the discharge across a circle including the source, you find it to be constant. The presence of the source therefore is felt everywhere, though the magnitude of the flow velocity becomes small far from the source.

Another example is a point vortex, where

$$\phi(x, y) = c\theta \tag{19.10}$$

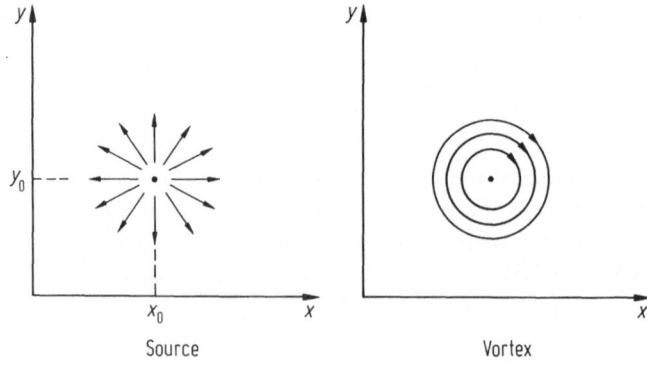

Fig. 19.2 Point source and point vortex.

in which θ is the angle in a polar coordinate system with its origin at the location of the vortex. Note that the potential equation is not valid in that point. The flow field is illustrated in Fig. 19.2. Again the flow velocity becomes smaller but not zero far away. Apparently, in an elliptic system, the influence of a disturbance is felt immediately and everywhere (compare with parabolic sytems, sections 7.1 and 15.6, and hyperbolic systems, section 15.2).

As there exist no characteristics, you cannot apply the rule for the number of boundary conditions given in section 15.2. Although there is a consistent mathematical theory of boundary conditions for elliptic equations, it is rather complicated, so you may be satisfied with the following argument of plausibility. A one-dimensional equivalent of an elliptic equation might look like

$$\frac{d^2\phi}{dx^2} = \lambda\phi$$

For such an equation, you need two boundary conditions. If λ is positive and large, you get exponentially growing solutions, which cannot be controlled very well by having the two boundary conditions on one side. It is better to have one condition on both sides of the region. In a sense, the term $\partial^2\phi/\partial y^2$ may behave in the same way as $\lambda\phi$, which means that you should have one boundary condition on each boundary, looking in the x-direction. However, as the equation is symmetric in x, y, the same argument is valid in the y-direction. Combining these you find that you must have a closed boundary all around the region of interest, with one boundary condition in each point of that boundary.

The boundary condition may be

(i) ϕ specified ("Dirichlet" condition);
(ii) $\partial\phi/\partial n$ specified, with n the direction normal to the boundary ("Neumann" condition);
(iii) a relation between the two.

Examples are given later.

19.4 Pressure

In many cases, you will not be interested in the potential, not even in the flow velocity, but rather in the forces exerted by the fluid on an obstacle. There are generally two kinds of forces: one due to pressure differences and one due to friction. The latter, however, cannot be found from potential flow as friction has explicitly been neglected. If the pressure forces are dominant, the potential-flow approximation can be useful. Substituting the definition of the potential into the Navier–Stokes equations (14.9) and (14.11) (with w, z replaced by v, y) using eq. (19.2) and omitting the viscosity terms, you find

$$-\frac{\partial}{\partial t}\left(\frac{\partial \phi}{\partial x}\right) + u\frac{\partial u}{\partial x} + v\frac{\partial v}{\partial x} + \frac{1}{\rho}\frac{\partial p}{\partial x} = 0$$

$$-\frac{\partial}{\partial t}\left(\frac{\partial \phi}{\partial y}\right) + u\frac{\partial u}{\partial y} + v\frac{\partial v}{\partial y} + \frac{1}{\rho}\frac{\partial p}{\partial y} + g = 0$$

(19.11)

Integrating these with respect to x and y gives

$$-\frac{\partial \phi}{\partial t} + \frac{1}{2}(u^2 + v^2) + \frac{p}{\rho} + gy = F(t)$$

(19.12)

where $F(t)$ is a constant of integration, which does not depend on x, y, but possibly on t. This function is unknown, but will be fixed if the pressure is fixed in one point. As you are always interested in pressure *differences*, the value of this reference pressure is not important and you may choose $F = 0$ if you like. Then, the pressure can be determined from the potential by

$$\frac{p}{\rho} = \frac{\partial \phi}{\partial t} - \frac{1}{2}\left\{\left(\frac{\partial \phi}{\partial x}\right)^2 + \left(\frac{\partial \phi}{\partial y}\right)^2\right\} - gy$$

(19.13)

You should be warned that this does not always give sensible results. Particularly, the notorious paradox of D'Alembert states that the total force (not the pressure in an individual point) on an obstacle in a steady potential flow will always be zero. On the other hand, for forces on obstacles in waves, the time derivative in eq. (19.13) is important and this may give quite reasonable results.

19.5 Exercises

1. The transport of a dissolved substance in water (compare section 14.1) is described by the 2-d convection-diffusion equation

$$u\frac{\partial c}{\partial x} + v\frac{\partial c}{\partial y} - D\left(\frac{\partial^2 c}{\partial x^2} + \frac{\partial^2 c}{\partial y^2}\right) = 0$$

Show that this is an elliptic equation. Hint: introduce auxiliary variables $p = \partial c/\partial x$, $q = \partial c/\partial y$ and produce two first-order differential equations for them. Then follow the procedure of section 19.3.

2. Check that eqs. (19.9) and (19.10) satisfy the potential equation. Show (by integrating the radial velocity component) that the discharge across any circle around a point source is constant.

3. You are supposed to compute the potential flow in a passage shown in Fig. 19.3. Which boundary conditions (in terms of velocity) would you impose in points A, . . . , G? How are these translated into boundary conditions for the potential? And for the stream function?

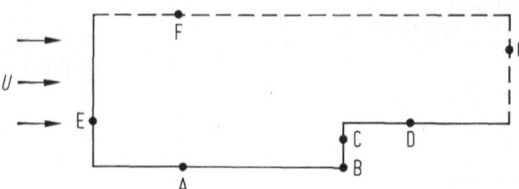

Fig. 19.3 Example of boundary conditions for potential flow.

Chapter 20
Finite-Difference Method for Potential Flow

20.1 Difference Equation

For a numerical solution of the potential equation, the region in the (x, y) plane is covered by a grid with sizes Δx, Δy (not necessarily equal); an example is given in Fig. 20.1. If you limit yourself to regular and rectangular grids, the boundaries of the region have to be represented by broken lines through the nearest grid points; the same difficulty was met in Chapter 18. Here, there is no question of staggered grids, as there is only one unknown. If you replace each term in the potential equation (19.4) by the straightforward finite-difference expression (there is not too much of a choice), you get for the special case of $\Delta x = \Delta y$:

$$\phi_{k+1,j} + \phi_{k-1,j} + \phi_{k,j+1} + \phi_{k,j-1} - 4\phi_{k,j} = 0 \tag{20.1}$$

If you mark the weighting coefficients for each grid point in Fig. 20.1, you obtain a "finite-difference molecule" connecting each grid point to its four direct neighbours. Each equation (20.1) contains five unknowns, so that the method is implicit and you will have to solve the set of such equations for all grid points together.

Equation (20.1) cannot be applied in boundary points as you would need grid points outside the region. However, in each boundary point a boundary condition is supposed to be available. It may be that the potential itself is given, so it is no longer unknown. Things get a little more complicated if the normal derivative is specified, for example at the right-hand boundary of Fig. 20.1:

$$\phi_{K,j} - \phi_{K-1,j} = \Delta x g_{K,j} \quad \text{(known)} \tag{20.2}$$

This produces an equation for the unknown boundary potential, so together with eq. (20.1), you end up with a system of exactly the right number of equations for all unknowns. A "mixed" boundary condition is handled the same way. Actually, it is a little more accurate to choose the grid such that the boundary is just halfway between points (K, j) and $(K - 1, j)$. Then eq. (20.2) is a central difference instead of a backward one, and the result will be more accurate. There is a way of handling boundaries that do not coincide with grid points (e.g. Mitchell 1977), but this is much more complicated.

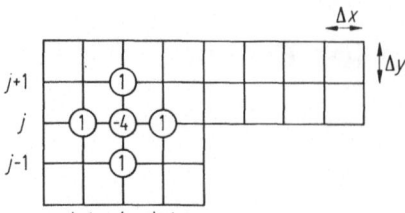

Fig. 20.1 Five-point "molecule".

The system of eq. (20.1) possibly with eq. (20.2), has a band structure, which you can see as follows. Suppose you number the grid points row by row, starting in the top left corner and ending in the right-hand corner of the lower line. There may be more efficient ways of numbering, but this is not essential for the argument. Then, looking at eq. (20.1), you see that each grid point is related only to its left and right neighbours, which have numbers one lower and higher, and to one point on the preceding and following rows, which have numbers lower or higher by the number of points in one row. Consequently, you get nonzero coefficients in the matrix on the main diagonal and its direct codiagonals, together with some locations at a distance of once the number of points per row. All other coefficients are zero. This is what is called a sparse matrix.

$$
\begin{pmatrix}
-4 & 1 & & & & 1 & & & & \\
1 & -4 & 1 & & & & 1 & & & \\
 & 1 & -4 & 1 & & & & 1 & & \\
 & & & & & & & & 1 & \\
1 & & & & & & & & & \\
 & 1 & & & & & 1 & -4 & 1 & \\
 & & 1 & & & & & 1 & -4 & 1 \\
 & & & 1 & & & & & 1 & -4 & 1
\end{pmatrix}
\Phi = 0 \qquad (20.3)
$$

If the rows do not have a constant number of grid points (as is the case in Fig. 20.1), the outer diagonals are not a constant distance from the main diagonal, so the matrix has a somewhat irregular "profile".

Solving such a system with band structure can be done with the standard Gauss elimination method, for which you can reference any good book on numerical methods. The advantage of the band structure is that you do not have to do eliminate the matrix entries outside the outer diagonals, as they are zero already. All coefficients inside the band will, however, be filled in with nonzero numbers during the process of elimination. This means that the required computer storage will be of the order $2mn$ if n is the total number of unknowns and $m \simeq n^{1/2}$ the (half) band width. The number of arithmetic operations, which is of the order of n^3 for a 'full' matrix, is now of the order $nm^2 \simeq n^2$.

For large systems, these numbers (i.e. computer storage and computer time) may get very large and you may consider iterative methods to solve the system. These are much more economic for storage as they use only the nonzero coefficients in the original system (there is no "fill-in"). However, the convergence of such methods is usually very slow, particularly for large systems. It may therefore be necessary to use rather sophisticated iterative methods. Again, you are referred to the specialized literature in this field.

20.2 Accuracy

In previous chapters, you have seen three ways of judging the accuracy of a numerical method:

(i) determine the truncation error;
(ii) find an expression for the numerical solution in a special but typical case and compare with the analytic solution;
(iii) repeat the actual computation with smaller grid size and observe any changes in the numerical results.

Unfortunately, in the case of elliptic equations the second approach can hardly be used as there are no representative (and sufficiently simple) solutions available. The other two methods can be used.

The truncation error of eq. (20.1) can be determined in a straightforward way by Taylor expansion with respect to the central point (k, j). You may check that the modified equation is

$$\frac{\partial^2 \phi}{\partial x^2} + \frac{\partial^2 \phi}{\partial y^2} = -\frac{1}{12} \Delta x^2 \frac{\partial^4 \phi}{\partial x^4} - \frac{1}{12} \Delta y^2 \frac{\partial^4 \phi}{\partial y^4} + \ldots \qquad (20.4)$$

The method is therefore of second order accuracy, i.e. if you reduce the grid size by a factor of 2, the error will decrease by a factor of 4. Note that the amount of work in case of such a reduction increases quite drastically: the number of equations gets larger by a factor of 4 and the total work (in case of a direct solution technique) even by a factor of 16. Moreover, the error is not distributed uniformly: eq. (20.4) shows that it is related to the 4th derivatives of the solution, which may get large, for example, near corners.

Equation (20.4) gives the error in the equation. For a smooth solution, you may assume that the error in the solution will also be of order Δx^2, but this is not true in the vicinity of corners.

The experimental method of judging the accuracy is illustrated in a simple example of potential flow in a channel with a step (Fig. 20.2). The problem is solved in terms of the stream function, for which the boundary conditions are shown in the figure.

Numerical results have been obtained for three grids (Fig. 20.3). The computed streamlines (lines of constant ψ) are not smooth. This is a result of the contour line

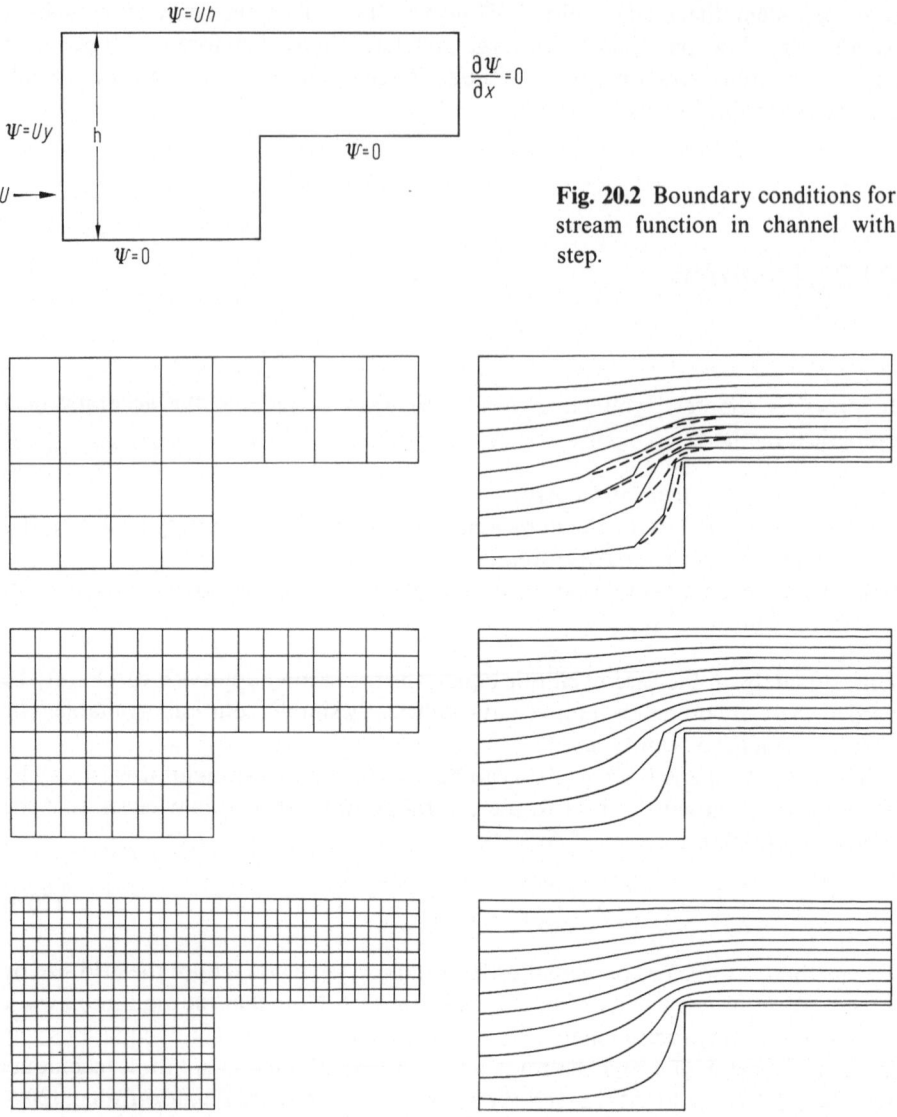

Fig. 20.2 Boundary conditions for stream function in channel with step.

Fig. 20.3 Grids and computed streamlines for 2, 4 or 8 points per step height.

program which must interpolate in some way between the grid values; in this case linearly.

The streamlines for the finest grid are superimposed on the coarsest to see the difference, which is noticeable but not very large. The difference between the medium and fine grid solutions is even smaller. If you consider the step height as a relevant length scale in this example, you may conclude that a grid size about 10 to 20% of this is sufficient for a reasonable accuracy.

20.3 Example

An object (e.g. a tunnel element being lowered to the bottom during construction) is submerged in water with waves. A longitudinal section is shown in Fig. 20.4. Suppose for simplicity that the situation is two-dimensional. You are asked to compute the hydrodynamic forces on the object.

For the purpose of illustration, a number of simplifications are introduced that may not be justified in actual applications:

– the water surface is assumed fixed;
– the oscillating flow is independent of depth in absence of the tunnel element;
– the flow is varying harmonically in time;
– the displacement of water particles is assumed to be small compared with the dimensions of the tunnel element; this means that flow separation does not play a significant part.

The first two assumptions imply that the length of the wave is large; the last that its amplitude is small.

With these assumptions, you may use the potential-flow formulation. The boundary conditions are:

$$x = 0 \quad \frac{\partial \phi}{\partial x} = -u_0 \sin \omega t$$

$$x = L \quad \frac{\partial \phi}{\partial x} = -u_0 \sin \omega t$$

$$z = 0 \quad \frac{\partial \phi}{\partial z} = 0$$

$$z = a \quad \frac{\partial \phi}{\partial z} = 0$$

$$\text{on the obstacle} \quad \frac{\partial \phi}{\partial n} = 0$$

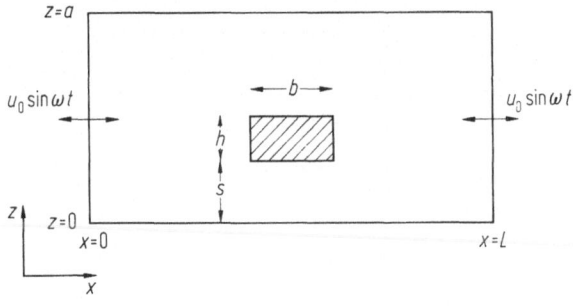

Fig. 20.4 Tunnel element in oscillating flow.

You will recognize that the entire solution gets harmonic in time, so the time-dependent part can be split off:

$$\phi(x, z, t) = u_0 \sin \omega t \, \phi'(x, z)$$

The boundary conditions then become very simple:

$$x = 0, \quad x = L \quad \frac{\partial \phi'}{\partial x} = -1$$

$$z = 0, \quad z = a \quad \frac{\partial \phi'}{\partial z} = 0$$

As the in- and outflow boundary conditions assume no influence of the obstacle, they must be far away from it. Some experiments with the numerical model have been done to find out the required distance. The numerical solution for ϕ' is straightforward using the five-point method of section 20.1. The following data have been used:

$$L/a = 2 \text{ or } 4$$
$$a/h = 3$$
$$b/h = 2$$
$$h/s = 2$$

Grid sizes varying from $a/12$ to $a/72$ have been used. Figure 20.5 gives the computed potential (differences between the various runs are difficult to see in this sort of figures).

The potential is not the quantity you have been looking for. The forces on the obstacle must be computed from the pressure, which in its turn can be found from

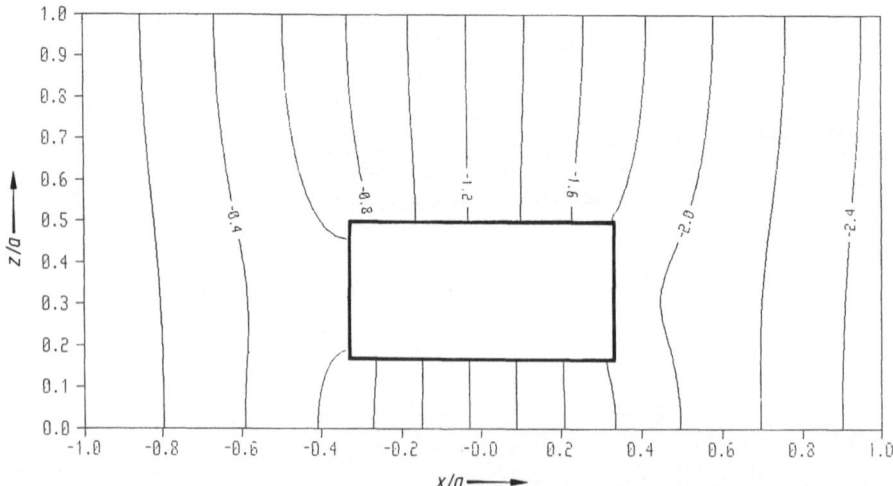

Fig. 20.5 Computed equi-potential lines.

eq. (19.13). The horizontal force F_x is then found by integrating the pressure over the left and right faces of the obstacle and subtracting; similarly for the vertical component.

As stated in the discussion of eq. (19.13), the horizontal force is determined by the time rate of change of the potential only; the other parts do not contribute to the net force. Therefore,

$$F_x = u_0 \omega \cos \omega t \, F'_x$$

For the vertical or lift force, the situation is different. The potential will be antisymmetric in x-direction with respect to the middle of the obstacle. This means that the term $\partial \phi / \partial t$, when integrated, gives a zero result (note that this is due to the special shape of the obstacle; it is not a general property). The quadratic part of eq. (19.13) results in a force

$$F_z = \tfrac{1}{2} u_0^2 (1 - \cos 2\omega t) F'_z$$

i.e. the lift force has a constant component and another one oscillating at the double frequency. The reason is that the flow direction does not matter as far as the vertical force is concerned.

Numerical results for the two force components as depending on the numerical grid size are as follows:

L/a	$\Delta x / \Delta z$	$a/\Delta x$	F'_x	F'_z
2	1	24	0.437	0.105
		48	0.450	0.107
		72	0.453	0.106
4	1	24	0.424	0.086
		48	0.442	0.096
		72	0.449	0.098
4	2	48	0.437	0.100

This shows that refining the grid gives converging results for the force components. The table also shows that a grid ratio $\Delta x / \Delta z = 2$ does have some influence. The location of the artificial in- and outflow boundaries has some effect as well, particularly on the lift component.

Generally, the lift component turns out to be rather sensitive to this sort of numerical parameters. This must be attributed to singularities in the velocity (which is the gradient of the potential) at the corners of the obstacle, which have not been taken into account. Although they are of a local nature, they have some influence on the integrated results. For more accurate results, they would have to be taken care of, but it is doubtful whether this would really pay, considering the other approximations involved.

20.4 Exercises

1. You could try to solve eq. (20.1) in a "marching" fashion, i.e. assume y-levels $j-1$ and j to be known and solve level $j+1$ from it. This would even be an explicit process. There is a difficulty with the boundary conditions (you would need two boundary conditions at $j=0$ to get started). Apart from that, show that the method would not be very useful as it is always unstable.

2. In the example of section 20.3, the hydrostatic part of the pressure (last term in eq. (19.13)) has not been mentioned. What is its effect on the horizontal and vertical force components?

Chapter 21
Finite-Element Method

21.1 Principle

If you want to compute flows in regions of a nicely rectangular form, there is no problem in using regular grids as in the previous chapter. By means of "telescoping grids" (Fig. 21.1), it is even possible to concentrate grid points in certain parts of the region where you want to have higher accuracy, for example because you expect the higher derivatives in the truncation error (eq. 20.4) to be large. However, for regions with a complicated shape such modifications are insufficient. What you would like to have is a grid that (i) is adapted to the boundaries, and (ii) has the possibility of local refinement for better accuracy. In recent years, such techniques have been developed in finite-difference form (see, e.g. Thompson et al., 1985). Another very powerful method is the finite-element method, of which a very simple variety is discussed here, based on a triangular "grid" (Fig. 21.2). There are several good books available, both for beginners and specialists (Pinder & Gray 1977, Taylor & Hughes 1981, Baker 1983, Chung 1978).

The region is first divided more or less arbitrarily into finite elements of triangular form. As you will suspect, the smaller the elements, the better the accuracy. Other variants use quadrilateral elements or even elements with curved sides. The basic approach is the same.

The second step is to represent the unknown function ϕ by a finite number of parameters. To explain the idea, let us first consider a function of one variable $\phi(x)$. This is approximated as a sum of standard functions $f_k(x)$:

$$\bar{\phi}(x) = \sum_{k=1}^{n} \bar{\phi}_k f_k(x) \tag{21.1}$$

You could choose polynomials or sine functions for these standard functions; however, the successful choice in the finite element method is a piecewise linear function, satisfying the following conditions:

(i) $f_k(x_j) = 0$ in all grid points (or nodes) x_j for which $k \neq j$;
(ii) $f_k(x_k) = 1$, i.e. each grid point has one standard function which equals unity there;
(iii) in between the grid points, all standard functions vary linearly.

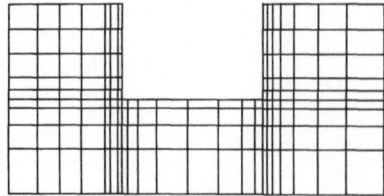

Fig. 21.1 Rectangular but non-uniform grid.

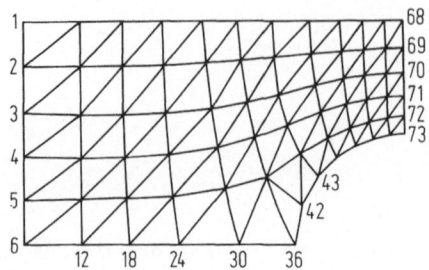

Fig. 21.2 Triangular finite-element grid.

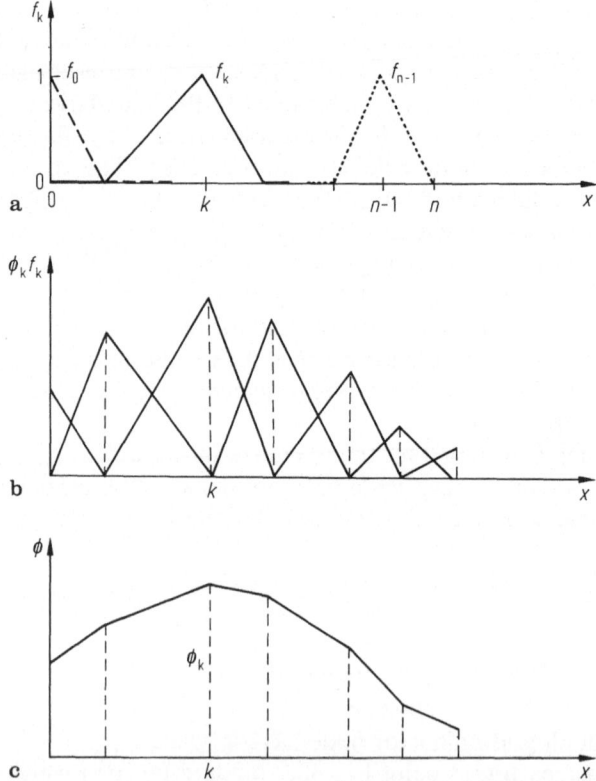

Fig. 21.3 Interpolation in one dimension; a. some standard functions; b. the separate terms of eq. (21.1); c. the interpolated function $\phi(x)$.

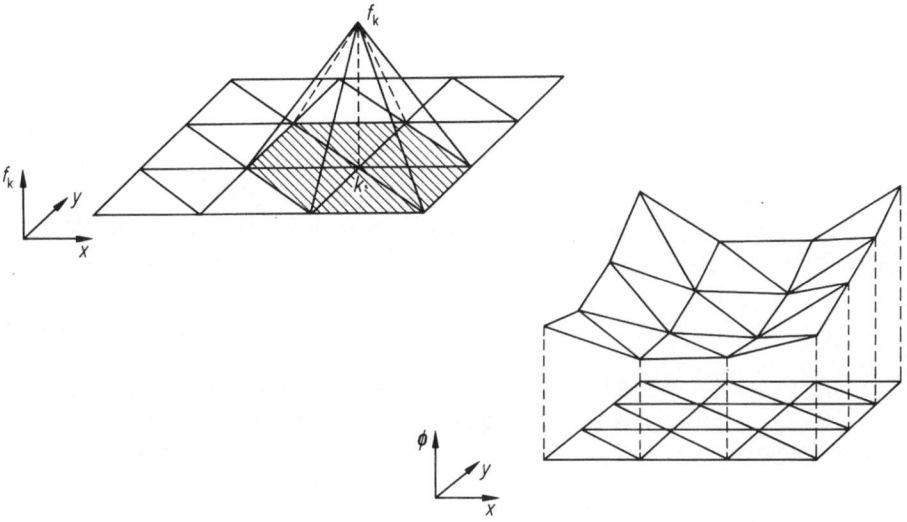

Fig. 21.4 Interpolation in two dimensions; a. a standard function; b. the interpolated function $\phi(x, y)$.

A few such functions are illustrated in Fig. 21.3(a). Note that the definition works on the boundaries as well. Each standard function is zero in a great part of the region, except within the elements next to its "own" node. The standard functions are continuous, but their first derivative is discontinuous in the nodes. Second derivatives do not exist.

In Fig. 21.3(b), the separate terms of eq. (21.1) are shown; if these are added together, you get a continuous curve consisting of straight-line pieces (Fig. 21.3(c)). This is the type of function with which finite-element methods work. In more refined variants, the functions may be piecewise quadratic or cubic. You will note that the parameters ϕ_k are the values of the function in the grid points. The standard functions do nothing more than interpolate between these grid values. This is why they are usually referred to as interpolation functions.

In two dimensions, the idea is exactly the same. In particular, the three rules for interpolation functions are used without any change. With a little thought, you will agree that this leads to functions of the form illustrated in Fig. 21.4(a) and the interpolated function looks like Fig. 21.4(b). Again, the values ϕ_k are the nodal values.

21.2 The Galerkin Method

By means of the discretization procedure of the preceding section, you have now a continuous function characterized by the n nodal values ϕ_k. These must be determined such that the differential equation and boundary conditions are satisfied. If you substitute the approximate solution (21.1) into the differential equation (19.4), the result will be some nonzero, but hopefully small, residual r:

$$\frac{\partial^2 \phi}{\partial x^2} + \frac{\partial^2 \phi}{\partial y^2} = r(x, y) \tag{21.2}$$

The nodal values should be determined such as to minimize the residual in some sense. One way to do this is to multiply eq. (21.2) with some weighting function $w(x, y)$, integrate over the region and require that this weighted-average residual is zero;

$$\iint r(x, y) w(x, y) \, dx \, dy = 0 \tag{21.3}$$

This is one equation with n nodal values as unknowns. You may now choose n different weighting functions to obtain n equations for the nodal values, which can then be solved. The remaining question is how to choose these weighting functions.

In the Galerkin method, you take the interpolation functions $f_k(x, y)$ for this purpose. They are exactly the right number and, due to their definition, are independent. Moreover, they are nonzero only in a small subregion. Equation (21.3) means that the average residual in all these small subregions is zero. It is plausible that it must then be small throughout the region, which is exactly what you want.

When elaborating eq. (21.3), you will meet a small difficulty in that the expression (21.1) is not differentiable twice. However, use can be made of Gauss' theorem (see any good book on calculus), which gives

$$\iint w \left(\frac{\partial^2 \phi}{\partial x^2} + \frac{\partial^2 \phi}{\partial y^2} \right) dx \, dy = - \iint \left(\frac{\partial \phi}{\partial x} \frac{\partial w}{\partial x} + \frac{\partial \phi}{\partial y} \frac{\partial w}{\partial y} \right) + \oint w \frac{\partial \phi}{\partial n} \, ds = 0 \tag{21.4}$$

in which the latter term is a contour integral involving the normal derivative. You notice that in the right-hand part of eq. (21.4) only first derivatives play a part and they are very well defined for eq. (21.1). Using this *weak form* of the equations is a rather essential step in finite-element methods.

Substituting eq. (21.1) into (21.4) gives

$$\sum_{k=1}^{n} \phi_k \iint \left(\frac{\partial f_j}{\partial x} \frac{\partial f_k}{\partial x} + \frac{\partial f_j}{\partial y} \frac{\partial f_k}{\partial y} \right) dx \, dy - \oint f_j \frac{\partial \phi}{\partial n} \, ds = 0 \tag{21.5}$$

which should be applied for all node numbers j where the potential is unknown. This excludes boundary points with Dirichlet boundary conditions, but includes all other boundary points. The integrals in eq. (21.5) are independent of the solution, but depend only on the geometry and the element mesh.

21.3 Boundary Conditions

In order to see how the boundary conditions are handled you may discern two types of conditions.

(i) Essential Boundary Condition

In this case, the potential itself is given in the boundary node j (Dirichlet condition). For such nodes, eq. (21.5) is not applied. For all other nodes, the weighting function f_k is zero in node j (see the properties of the interpolation functions), so there is no contribution to the contour integral in eq. (21.5) (there is a contribution to the surface integral, in which the known boundary value can be used).

(ii) Natural Boundary Condition

This involves the normal derivative:

$$\frac{\partial \phi}{\partial n} = g\phi + h \tag{21.6}$$

where g may be zero in case of a pure Neumann boundary condition. In this case, eq. (21.5) should be applied to the boundary point j as well and there results a contribution to the contour integral:

$$\sum_{k=1}^{n} \phi_k \oint f_j f_k \, g \, ds + \oint f_j h \, ds \tag{21.7}$$

The natural boundary conditions are not imposed very strictly, but in a weighted-average sense. Consequently, they will not be satisfied exactly. The finer the finite-element mesh, the more accurate the approximation of the differential equation and the natural boundary conditions will be.

21.4 Comparison with Finite-Difference Method

Collecting the boundary contributions and introducing a short-hand notation for the geometric quantities, you get a system of linear equations, from which the nodal values can be solved:

$$\sum_{k=1}^{n} a_{kj} \phi_k = b_j \qquad (j=1, \ldots, n) \tag{21.8}$$

where

$$a_{kj} = \iint \left(\frac{\partial f_j}{\partial x} \frac{\partial f_k}{\partial x} + \frac{\partial f_j}{\partial y} \frac{\partial f_k}{\partial y} \right) dx\, dy - \oint g f_j f_k\, d\,s \tag{21.9}$$

$$b_j = \oint f_j h\, ds \tag{21.10}$$

A certain node is connected to those neighbouring nodes that share an element with it. For all other nodes, the coefficient a_{kj} from eq. (21.9) is zero as the functions f_k and f_j do not have any region in common where they are both nonzero. Therefore, the matrix a_{kj} is a sparse one. The structure of this system depends on the way of numbering the nodes. If you number them row-wise (as in section 20.1), you get a band structure that resembles that of the finite-difference method very much. The band width is related to the number of nodes in one row. The same type of solution techniques can be used.

The construction of the finite-element equations may seem rather complicated to you. However, in practice you do not have to bother about all the details, as there exist many standard computer programs solving such matters as:

- generation of a finite-element mesh, where you only have to provide global indications how you want to have it;
- computation of the integral contributions (21.9) and (21.10) element-by-element and assembly into the matrix of eq. (21.8);
- solution of the system (21.8);
- graphical representation of the results.

Just in case you want to develop your own program, the relevant formulae for the linear triangular elements are given in Appendix 2.

In order to see the relation with finite-difference methods, you can take the special case of a rectangular grid, divided into triangles by diagonals (Fig. 21.5). A particular node is then related only to six surrounding nodes. The values of the matrix elements a_{kj} are indicated in the figure. Comparing this with Fig. 20.1, you see that they are identical. Therefore, there is not so much difference between the two methods as is sometimes suggested. You could consider the finite-element

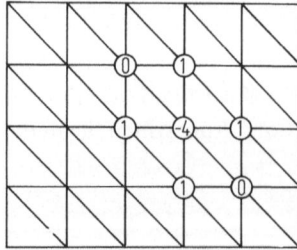

Fig. 21.5 Finite-element coefficients on a rectangular grid.

method as a general method to handle irregular grids, of which the standard finite-difference method is a special case. Anyway, the close agreement indicates that both methods will be of comparable accuracy if they have roughly the same grid size.

In the finite-element method, there are two ways to improve the accuracy if you are not satisfied with it. First of all, you can use a finer grid just as in the finite-difference case. On the other hand, you may leave the grid size as it is, but use higher-order interpolation functions. See for example Chung (1978) for a further discussion. In practice, the choice may often be dictated by the standard computer program you are using. If it does not have the feature of higher-order elements, only the former choice is left to you.

21.5 Groundwater Flow

The flow of groundwater in the soil was discussed in Chapter 7 for the special case that the flow is mainly horizontal. This is no longer true if you consider flow around constructions in the soil, such as foundations, sheet piles, etc. The dynamic equation for flow in a porous medium has been proposed by Darcy and it expresses that there is an equilibrium between the pressure gradient and the net resistance of the soil particles to the flow (see, e.g., Verruijt, 1970):

$$u = -k\frac{\partial \phi}{\partial x} \qquad v = -k\frac{\partial \phi}{\partial y} \tag{21.11}$$

in which k is a permeability coefficient as before. The "pressure" $\phi(x, y)$ is the height of the water column that would be measured in a vertical tube with the lower end positioned at the point (x, y). It is not necessarily equal to the ground water level above that point (as was assumed in Chapter 7). In addition to these dynamic equations, there is the usual mass-balance equation for two- dimensional flow

$$\frac{\partial u}{\partial x} + \frac{\partial v}{\partial y} = 0 \tag{21.12}$$

where it is assumed that both water and soil are incompressible (this excludes consolidation effects such as discussed in Chapter 10).

Substitution of eq. (21.11) into eq. (21.12) gives

$$\frac{\partial}{\partial x}\left(k\frac{\partial \phi}{\partial x}\right) + \frac{\partial}{\partial y}\left(k\frac{\partial \phi}{\partial y}\right) = 0 \tag{21.13}$$

If k is a constant, this reduces to the potential equation (19.4). If it is not, the equation is still elliptic.

There is apparently a great deal of similarity with potential flow (Chapter 19); yet there are some important differences. First of all, it *follows* from eq. (21.11) that the flow is irrotational (at least if k is a constant) rather than that it is an assumption, nor is there any assumption that the flow is frictionless. On the contrary, eq. (21.11) indicates that friction is predominant. Secondly, the interpretation of the potential is different: here it is identified with the pressure; in the case of potential flow there is a complicated relation (19.13) between the two. You should keep these differences carefully in mind.

All numerical methods discussed for potential flow can obviously also be used for groundwater flow. Here, an example is given in which the finite-element method is applied.

Figure 21.6 illustrates a concrete dam with sheet piles. The question is how much water will seep underneath the dam. It is supposed that the soil is uniform and that a relatively impermeable layer is present at a certain depth; this constitutes the lower boundary EF of the computational region. The vertical boundaries AF and DE are artificial and their location has to be considered carefully. In case A, they are located at a distance of twice the horizontal size of the dam; in case B and C at four times this size.

The boundary conditions are also indicated in Fig. 21.6. On AB there is a layer of water with depth h; if you remember the significance of the potential in this case, it means that $\phi = h$ on that boundary. On the contour BC of the structure, the normal component of the flow velocity must vanish, which means $\partial\phi/\partial n = 0$. The same applies to the impermeable layer EF. On CD, it is assumed that some water may just reach the surface, which means that $\phi = 0$ (this is an approximate condition). Finally, the two artificial boundaries must be treated. If they are far enough from the structure, you could assume that there is no flow at all, which means $\partial\phi/\partial n = 0$. As an alternative, you could assume that, although some flow may still be present, the pressure distribution is the same as in the static case, i.e. $\phi = h$ on AF and $\phi = 0$ on DE (note that the hydrostatic part has been neglected; this is acceptable as it is present on both sides and therefore does not produce any flow). The latter condition is evidently less restrictive and has been used in this example.

The size of the finite elements is determined by the expected variation of the potential. This will be greatest near the tip of the sheet wall. Using the finite-element technique, you have the opportunity to use a non-uniform mesh. Figure 21.7 gives some meshes generated by a commercial finite-element package

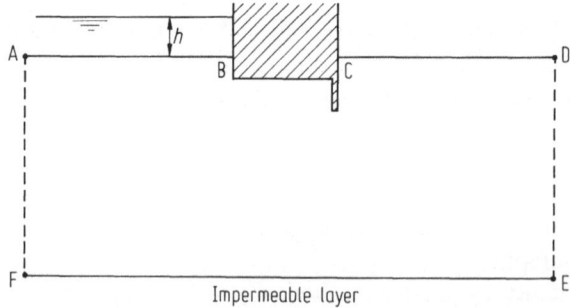

Fig. 21.6 Groundwater flow under a concrete dam.

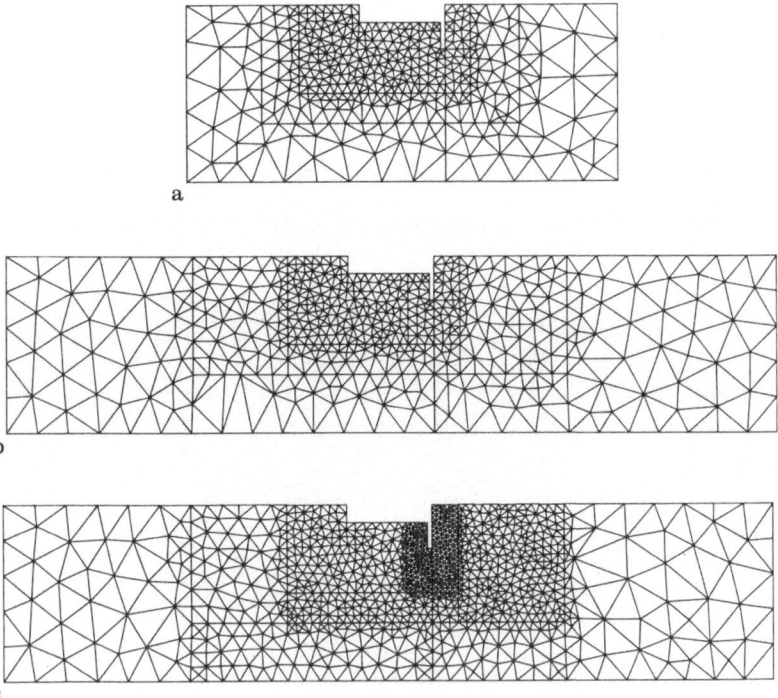

Fig. 21.7 Finite element meshes for cases A, B and C.

under control of some overall directives by the user. Zones can be distinguished in which mesh sizes of, respectively, about 1, 0.5 and 0.25 times the subsoil depth of the dam (which is 2 m in the example) have been used. In case C, a zone near the tip of the sheet wall has been further refined. In case B, the region has been extended by some very coarsely meshed areas to investigate the influence of the boundary location. Apart from this, cases A and B are similar. The automatic mesh generating program produces a more or less smooth transition between the various zones; the mesh itself is not quite regular. All elements are triangular with linear interpolation functions (as discussed in this chapter).

The computed potential is shown in Fig. 21.8. for cases A and B. There is a clear influence of the artificial boundaries which are too close in case A. It is assumed that case B is good enough (actually, to make sure you would have to run another case with an even larger computational domain). The result of case C is so close to that of B that it is not separately shown. The flow pattern is represented in Fig. 21.9 for case B. In each element, one velocity vector is shown (the velocity will be constant over an element due to the linear approximation of the potential). Corresponding pictures for the other cases are not shown as it is very difficult to tell the difference.

The differences can be seen somewhat more clearly by looking at the inflow into the soil left of the dam and the outflow right of it (Fig. 21.10) and to the flow in a

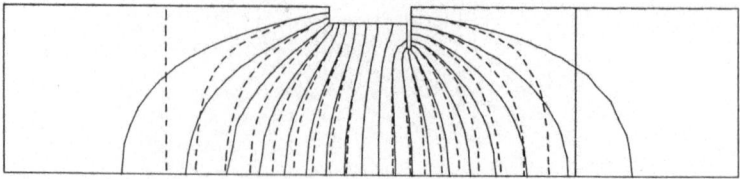

Fig. 21.8 Equi-potential lines for cases A (dashed) and B (drawn).

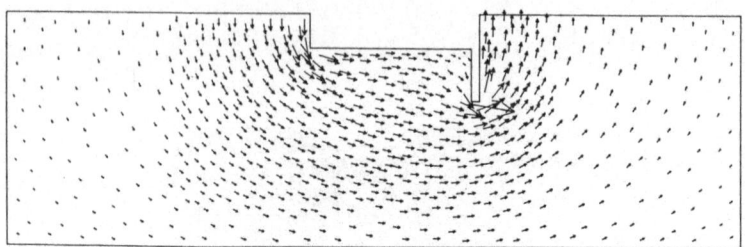

Fig. 21.9 Computed flow pattern for case B.

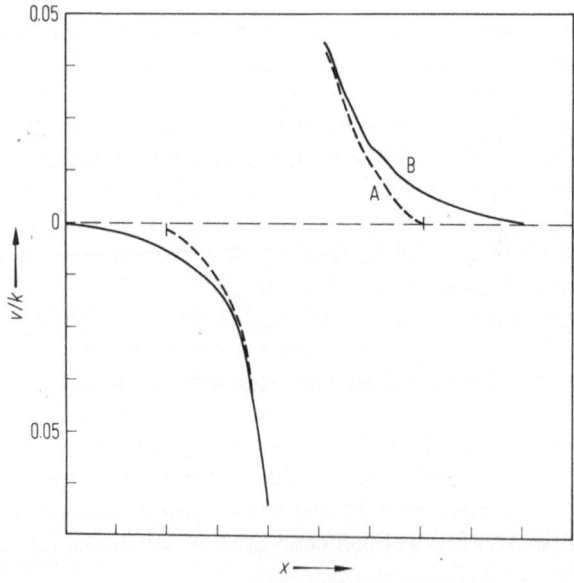

Fig. 21.10 Inflow and outflow across the top of the soil.

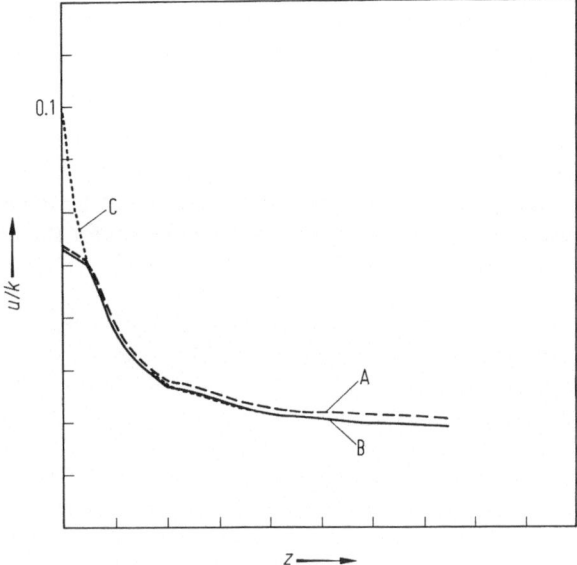

Fig. 21.11 Flow under the sheet wall.

vertical section below the sheet wall (Fig. 21.11). In both figures, u/k or v/k is given. The former figure confirms the conclusion that the boundaries in case A are too close. The result of case C is very near that of B. Figure 21.11 shows very strong gradients near the tip of the sheet wall, as could be expected. This is the only location where case C gives a significant difference.

Finally, the integrated in- and outflow rates across all boundaries are given to check the continuity and get a quantitative indication of accuracy.

Boundary	Case A	Case B	Case C
AF	0.173	0.037	0.037
AB	0.395	0.505	0.500
total inflow	0.468	0.542	0.537
flow under sheet wall	0.517	0.493	0.509
CD	0.356	0.512	0.501
DE	0.196	0.041	0.042
total outflow	0.452	0.553	0.543

For each model, the three numbers for inflow, flow under the sheet wall, and outflow should be equal. This turns out not to be true. The reason is that the sharp corners introduce local disturbances which are not important for the overall results, but show up in a detailed comparison. The results of cases B and C agree quite well. It may therefore be concluded that case B is satisfactory for all practical purposes. Case A, however, though it looks quite plausible in Fig. 21.7, is not good enough; its boundaries are too close.

21.6 Exercises

1. Show that the system (21.11), (21.12) is elliptic, i.e. that there are no real characteristics.
2. For groundwater flow as well, a stream function can be defined. Derive the differential equation for it. Compare with the equation for the potential (21.13).

Appendix 1
Long Waves

A1.1 Mathematical Formulation for Rivers

The equations describing long waves in the sea or in rivers can be obtained by integrating the general hydrodynamic equations over the depth or over a river cross-section (cf. e.g. Jansen, 1979). Here, a more heuristic derivation is given. For a "slice" of a river (Fig. A1.1), the mass balance gives in a straightforward way:

$$B\frac{\partial h}{\partial t} + \frac{\partial Q}{\partial x} = 0 \qquad\qquad (A1.1)$$

where h = waterlevel relative to a horizontal reference level;

$\qquad Q$ = discharge

$\qquad B$ = width at waterlevel.

In the same way, a momentum balance can be set up for the control volume. You should keep in mind that the momentum in the volume can change not only by in- or outflow (convection), but also by net forces acting on the volume of water.

The momentum in the volume is $\rho Q \Delta x$ (with ρ the density), so the rate of change is $\rho \Delta x\, \partial Q/\partial t$. The volume of water passing a cross-section per unit time is Q; the momentum associated with it is $\rho u = \rho Q/A_s$ and the momentum flux is the product of the two $\rho Q^2/A_s$. The difference of the momentum fluxes into and out of the volume is

$$\Delta x\, \partial/\partial x(\rho Q^2/A_s)$$

where A_s is the part of the cross-section that contributes to the flow. This may be the full cross-section, but it will be less if there are flood plains with obstructions such as groynes that essentially serve for storage only. Such regions do contribute to the equation of continuity, but not to the momentum equation.

The net force acting on the control volume consists of pressure and bottom friction. The former is illustrated in Fig. A1.2. As the flow is of the boundary-layer type (section 14.1), the pressure distribution will be hydrostatic; however, due to the slope of the water surface, there is a pressure difference between the two sides. If for simplicity, you assume a rectangular cross-section with width B_s, the total pressure

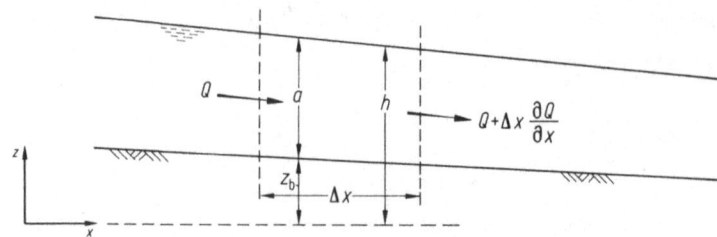

Fig. A1.1 Control volume for long-wave equations.

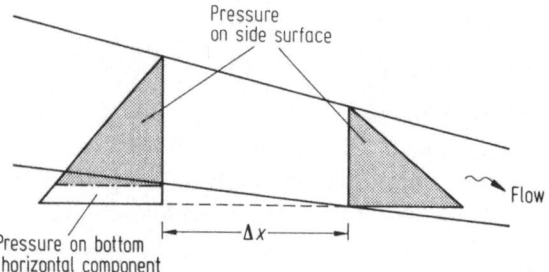

Fig. A1.2 Pressure differences.

on the left is $\frac{1}{2}\rho g a^2 B_s$, where a is the water depth (do not confuse with the water level h), similarly on the right side. Do not forget that a sloping bottom also produces a (component of the) pressure in the longitudinal direction; over a length Δx

$$- \rho g a \Delta x B_s\, \partial z_0/\partial x$$

where z_0 is the bottom level. The minus sign is included because the bottom slope is negative in Fig. A1.2. Collecting all pressure terms, you get

$$- \rho g a B_s\, \partial a/\partial x\, \Delta x - \rho g a B_s\, \partial z_0/\partial x\, \Delta x = - \rho g A_s \Delta x\, \partial h/\partial x$$

Finally, the bottom friction is $- B_s \Delta x\, \tau_b$. The bottom shear stress in a turbulent flow is proportional to the velocity squared, with an empirical frictional coefficient c_f:

$$\tau_b = \rho c_f u|u|$$

The absolute sign is included to make sure that the friction works in a direction opposite to the mean flow.

Now you can collect all terms of the momentum balance, divide by $\rho \Delta x$, and obtain

$$\frac{\partial Q}{\partial t} + \frac{\partial}{\partial x}\left(\frac{Q^2}{A_s}\right) + g A_s \frac{\partial h}{\partial x} + c_f B_s \frac{Q|Q|}{A_s^2} = 0 \tag{A1.2}$$

Equations (A1.1) and (A1.2), with a and Q as variables, together describe long waves in rivers or canals; they are often called the Saint Venant equations.

You can use depth a and velocity u as variables instead; to that end eq. (A1.1) with $Q = B_s au$ is reworked. Assuming B and B_s to be equal and independent of x (prismatic channel) this gives

$$\frac{\partial h}{\partial t} + \frac{\partial}{\partial x}(au) = 0 \tag{A1.3}$$

The same substitution in the first two terms of eq. (A1.2) gives

$$\frac{\partial Q}{\partial t} + \frac{\partial}{\partial x}\left(\frac{Q^2}{A_s}\right) = \frac{\partial}{\partial t}(A_s u) + \frac{\partial}{\partial x}(Qu) = u\left(\frac{\partial A_s}{\partial t} + \frac{\partial Q}{\partial x}\right) +$$

$$+ A_s\left(\frac{\partial u}{\partial t} + u\frac{\partial u}{\partial x}\right) = u(B_s - B)\frac{\partial h}{\partial t} + A_s\left(\frac{\partial u}{\partial t} + u\frac{\partial u}{\partial x}\right)$$

For the special case that $B = B_s$, you get (after division by A_s) the familiar form

$$\frac{\partial u}{\partial t} + u\frac{\partial u}{\partial x} + g\frac{\partial h}{\partial x} + c_f u|u|/a = 0 \tag{A1.4}$$

Equations (A1.3) and (A1.4) can be used as alternatives for eqs. (A1.1) and (A1.2); however, please note that they are not quite general.

A1.2 Mathematical Formulation in Two Dimensions

The flow in a shallow sea or a wide river is described in the same way, but the direction of the flow is no longer fixed. Momentum equations are needed in the two horizontal directions. The equation of continuity (mass balance) is now obtained by considering a fluid column of dimensions Δx, Δy:

$$\frac{\partial h}{\partial t} + \frac{\partial}{\partial x}(au) + \frac{\partial}{\partial y}(av) = 0 \tag{A1.5}$$

The momentum equations in the x and y directions are found to be

$$\frac{\partial u}{\partial t} + u\frac{\partial u}{\partial x} + v\frac{\partial u}{\partial y} - fv + g\frac{\partial h}{\partial x} + \tau_{bx}/\rho = 0 \tag{A1.6}$$

$$\frac{\partial v}{\partial t} + u\frac{\partial v}{\partial x} + v\frac{\partial v}{\partial y} + fu + g\frac{\partial h}{\partial y} + \tau_{by}/\rho = 0 \tag{A1.7}$$

There is one term new compared to the one-dimensional case: $-fv$ in the x-equation and $+fu$ in the y-equation. With a little thought you will see that this represents an acceleration in a direction normal to the flow direction; it is due to the rotation of the earth and is called the Coriolis acceleration. The coefficient is defined as

$$f = 2\Omega \sin \phi \tag{A1.8}$$

where Ω is the angular velocity of the earth's rotation and ϕ the geographical latitude.

In eqs. (A1.6) and (A1.7), the bottom friction is supposed to be proportional to the magnitude of the (depth-averaged) flow velocity, and directed opposite to it:

$$(\tau_{bx}, \tau_{by}) = \rho c_f (u^2 + v^2)^{1/2} (u, v) \tag{A1.9}$$

The properties of the two-dimensional Saint–Venant equations are similar to the one-dimensional case, discussed in the following sections.

A1.3 Characteristics

To find the characteristics, you apply the same method as in chapter 15, i.e. you look for some combination of the variables that will be conserved. As there are now derivatives of a involved in eq. (A1.3), it is better to choose the combination $f = ma + nu$. Using the formulation of eqs. (A1.3) and (A1.4), multiplying them by m and n respectively and adding them, you find

$$m \frac{\partial a}{\partial t} + mu \frac{\partial a}{\partial x} + ng \frac{\partial a}{\partial x} + n \frac{\partial u}{\partial t} + nu \frac{\partial u}{\partial x} + ma \frac{\partial u}{\partial x} + \ldots = 0 \tag{A1.10}$$

Now, in order to have the quantity f conserved you should find something of the form

$$\frac{\partial f}{\partial t} + c \frac{\partial f}{\partial x} + \ldots = 0$$

or

$$m \frac{\partial a}{\partial t} + cm \frac{\partial a}{\partial x} + n \frac{\partial u}{\partial t} + cn \frac{\partial u}{\partial x} + \ldots = 0 \tag{A1.11}$$

in which c is the celerity (unknown as well) of the characteristic. You find this to be possible if

$$mu + ng = cm \qquad nu + ma = cn$$

so

$$c = u \pm (ga)^{1/2} \tag{A1.12}$$

and

$$m = gn/(c - u)$$

The value of n is arbitrary and can be chosen to be unity. You find again that there are two characteristics in each point. Along each characteristic, you have from eq. (A1.11)

$$\frac{du}{dt} \pm (g/a)^{1/2}\frac{da}{dt} + \ldots = 0$$

which can be integrated to

$$\frac{d}{dt}(u \pm 2(ga)^{1/2}) + \ldots = 0 \tag{A1.13}$$

The dots indicate lower-order terms which may cause damping of the waves. They contain in particular the friction terms. The Riemann invariants are now

$$f = u + 2(ga)^{1/2} \qquad F = u - 2(ga)^{1/2} \tag{A1.14}$$

Using these, you can do exactly the same construction as in section 15.2.

Even these results are not quite general. You may derive as an exercise from eqs. (A1.1) and (A1.2) (by first differentiating all terms as far as possible and looking for a combination of a and Q to be conserved) that the general characteristics for the one-dimensional long-wave equations are:

$$c = u \pm \{gA_s/B + (1 - B_s/B)u^2\}^{1/2} \tag{A1.15}$$

A1.4 Linearization

For the purpose of analysis, you want the equations to be linear. This can be obtained by considering small disturbances on an equilibrium flow situation:

$$h = h_0 + h' \text{ etc.}$$

where h' is assumed small. Substituting this in the equation of continuity, you get (keep in mind that the surface width B is a function of depth):

$$\left\{ B(a_0) + a'\frac{\partial B}{\partial a} \right\}\left\{ \frac{\partial h_0}{\partial t} + \frac{\partial a'}{\partial t} \right\} + \frac{\partial Q_0}{\partial x} + \frac{\partial Q'}{\partial x} = 0$$

In the reference situation time and space derivatives are zero. You may neglect products of small quantities, so retain only linear terms:

$$B_0\frac{\partial a'}{\partial t} + \frac{\partial Q'}{\partial x} = 0 \tag{A1.16}$$

The same process is applied to the momentum equation. The algebra is a little more complicated, but you can check that, e.g., the convection term gives

$$\frac{\partial}{\partial x}(Q^2/A_s) = \frac{\partial}{\partial x}\left\{(Q_0 + Q')^2 \middle/ \left(A_{so} + \frac{\partial A_s}{\partial a}a'\right)\right\}$$

$$= \frac{\partial}{\partial x}\left\{(Q_0^2 + 2Q_0 Q')\left(1 - \frac{1}{A_{so}}\frac{\partial A_s}{\partial a}a'\right) \middle/ A_{so}\right\}$$

$$= \frac{\partial}{\partial x}(Q_0^2/A_{so}) + \frac{\partial}{\partial x}\left(2\frac{Q_0}{A_{so}}Q' - \frac{Q_0^2}{A_{so}^2}B_{so}a'\right)$$

Another one that needs careful attention is the bottom friction term (assume c_f constant, although this is not essential):

$$B_s c_f Q|Q|/A_s^2 = c_f\left\{\frac{B_s}{A_s^2} + a'\frac{\partial}{\partial a}\left(\frac{B_s}{A_s^2}\right)\right\}\{Q_0|Q_0| + 2|Q_0|Q'\}$$

$$= c_f B_{so}Q_0|Q_0|/A_{so}^2 + 2c_f B_{so}|Q_0|Q'/A_{so}^2 + c_f Q_0|Q_0|\frac{\partial}{\partial a}\left(\frac{B_s}{A_s^2}\right)a'$$

The first term balances the bottom slope in the uniform base flow. The linear terms with a' are collected:

$$\left\{gB_{so}\frac{\partial z_b}{\partial x} + c_f Q_0|Q_0|\frac{\partial}{\partial a}\left(\frac{B_s}{A_s^2}\right)\right\}a' =$$

$$\left\{-\frac{B_{so}^2}{A_{so}^3}c_f Q_0|Q_0| + c_f Q_0|Q_0|\frac{\partial}{\partial a}\left(\frac{B_s}{A_s^2}\right)\right\}a'$$

Before putting everything together, it is easier if the variables are made dimensionless by introducing

$$a'' = a'/a_0, \qquad Q'' = Q'/Q_0$$

$$t'' = t/T, \qquad x'' = x/(gA_{so}/B_0)^{1/2}T$$

where T is the wave period. The linearized equations can then be written as (dropping all the double primes):

$$\frac{\partial a}{\partial t} + bF\frac{\partial Q}{\partial x} = 0 \tag{A1.17}$$

$$\frac{\partial Q}{\partial t} + 2F\frac{\partial Q}{\partial x} + \left(\frac{1}{bF} - \frac{F}{b_1}\right)\frac{\partial a}{\partial x} + K(2Q - \gamma a) = 0 \tag{A1.18}$$

In these equations, some dimensionless parameters occur, that apparently determine what happens to the waves. You meet first of all some geometric parameters, depending on the shape of the cross-section. For a schematic profile, shown in Fig. A1.3, they can be computed

$$b = A_{so}/B_0 a_0 = B_{so}/B_0$$

$$b_1 = A_{so}/B_{so}a_0 = 1$$

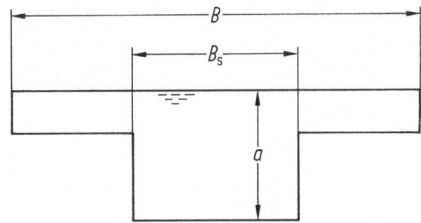

Fig. A1.3 Schematic rectangular cross-section.

$$\gamma = B_{s0} a_0 / A_{s0} - \frac{A_{s0} a_0}{B_{s0}} \frac{\partial}{\partial a} \left(\frac{B_s}{A_s^2} \right) = 1 + 2A_{s0}^2 a_0 / B_{s0}^2 a_0^3 = 3$$

In addition, there are two dynamic parameters: the Froude number

$$F = u_0 / (g A_{s0} / B_0)^{1/2}$$

and the friction parameter, mentioned in chapter 15, but given here in its complete form:

$$K = |u_0| T c_f / a_0$$

The Froude number is usually not very large (say, up to 0.4). However, depending on the wave period and the bottom friction, K may vary between 0 and 1000.

There is a complication if you consider oscillating tidal flow, because then the flow disturbance is not small compared with the base flow. For most of the terms, this is not very serious and you may take $u_0 = 0$. However, this is not allowed in the friction term, as it would disappear completely. There is an alternative way (attributed to the physicist Lorentz) of linearizing the friction term so that, averaged over a tidal cycle, it dissipates the correct amount of energy. This leads to an alternative definition

$$K = \frac{4}{3\pi} |\hat{u}| T c_f / a_0$$

to be used if the oscillating velocity component $\hat{u}$ exceeds, say, twice the reference flow velocity u_0.

A1.5 Wave Propagation

The behaviour of long waves can now be studied by solving eqs. (A1.17) and (A1.18). Suppose that a periodic fluctuation with period T is imposed at $x = 0$. In dimensionless form:

$$a(0, t) = H \exp(-2\pi i t)$$

If you have a more complicated boundary condition in reality, it can be decomposed into similar terms by developing it into a Fourier series. Each term can be considered separately and you can add the contributions afterwards due to the linearity of the equations. As you may expect, you will find a solution

$$a(x, t) = H \exp(\lambda x - 2\pi i t)$$

$$Q(x, t) = Q \exp(\lambda x - 2\pi i t)$$

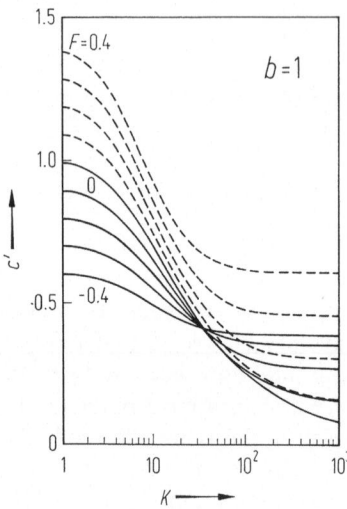

Fig. A1.4 Dimensionless wave speed ($b = 1$) Drawn lines: wave travelling upstream, dashed lines: wave travelling downstream.

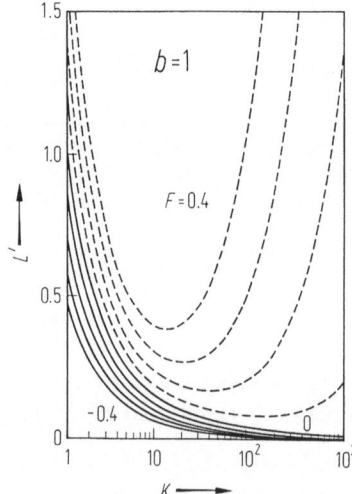

Fig. A1.5 Dimensionless damping length ($b = 1$).

You can solve the complex numbers λ and Q by substituting into the linearized equations of motion. This gives two homogeneous equations for H and Q, which have a nonzero solution only if the determinant of coefficients equals zero. This leads to two possible values of λ, from which you take the one with negative real part, because the other does not behave nicely at infinity. The algebra is tedious but straightforward; it is sufficient to give the result in Figs. A1.4 ... A1.7. The dimensionless quantities are defined in Section 15.5.

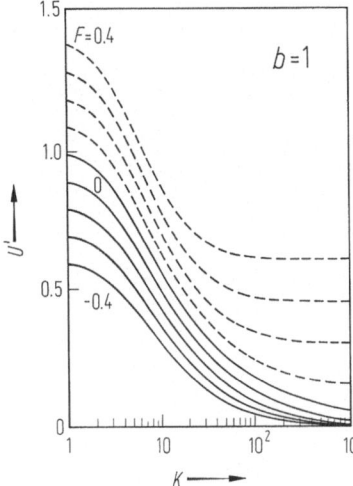

Fig. A1.6 Dimensionless amplitude of discharge $(b=1)$.

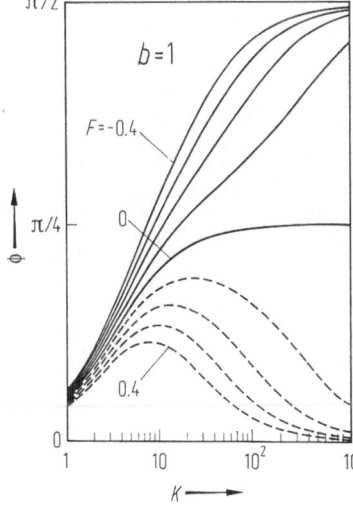

Fig. A.17 Phase angle between discharge and water level $(b=1)$.

You can observe some interesting points in these figures. If K is very small, which means that there is almost no influence of bottom friction (or a very small wave period), you find (taking the direction of propagation into account)

$$c' = F \pm 1$$

$$Q = B_s Hc \quad \text{(dimensional)}$$

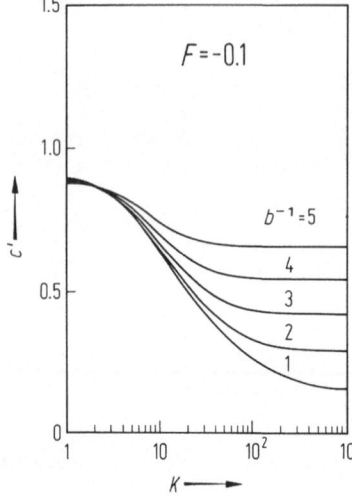

Fig. A1.8 Wave speed for various width ratios $b^{-1} = 1, \ldots, 5$.

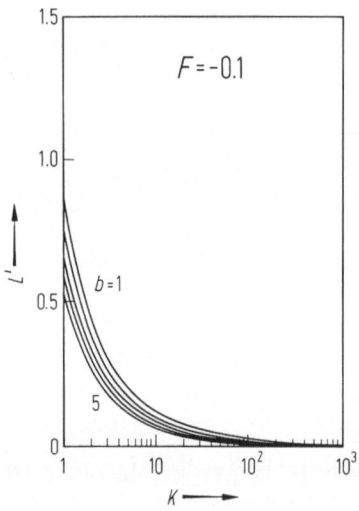

Fig. A.19 Damping length for various width ratios.

together with $L' \to \infty$ (no damping) and $\phi = 0$. This agrees exactly with the theory of characteristics (eqs. A1.12, A1.14). The latter may not be obvious, but if you introduce small disturbances into eq. (A1.14), you get

$$u' = (g/a_0)^{1/2} a'$$

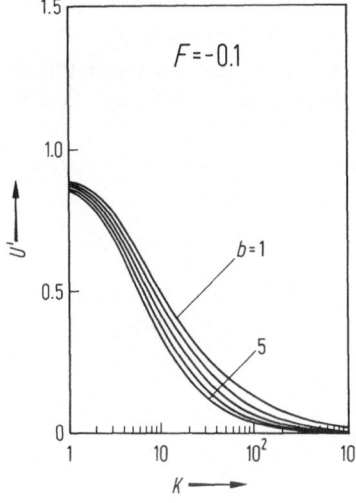

Fig. A1.10 Discharge amplitude for various width ratios.

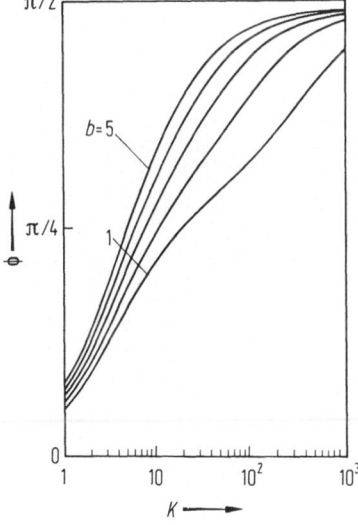

Fig. A1.11 Phase shift for various width ratios.

so

$$Q' = B_s(a_0 u' + u_0 a') = B_s\{u_0 + (ga_0)^{1/2}\}a' = B_s c a'$$

At the other extreme, if K gets very great, you may check that waves travelling upstream can no longer exist: their damping length goes to zero, so they are damped immediately. Waves travelling downstream with the base flow have a wave speed $c' = 1.5F$ and no wave damping, which corresponds to a kinematic wave. You see that there is a gradual transition between these two extreme wave types, when K increases.

For one value of the Froude number, the influence of the stream-to-storage width ratio b is shown. As it increases, the wave speed increases as well, the damping length decreases (so you have a stronger damping), the discharge amplitude decreases and the phase shift increases. The effects can be quite significant.

Appendix 2
Linear Triangular Finite Elements

The formulation of a finite-element method for the potential equation based on triangular grids and linear interpolation functions is given in eqs. (21.8) ... (21.10). The coefficients are usually split into elementwise contributions, which can be computed separately and added to the relevant entries in the matrix and right-hand side vector:

$$a_{kj} = \sum_{e=1}^{m} a_{kje} \qquad (A2.1)$$

$$b_j = \sum_{e=1}^{m} b_{je} \qquad (A2.2)$$

where e is the element number. Node and element numbers can be issued independently but you need a relationship between them, stating which node numbers belong to an element and, conversely, which elements are connected to a certain node. Most finite-element programs do this for you. The expressions for the elementwise contributions are:

$$a_{kje} = \iint_e \left(\frac{\partial f_j}{\partial x} \frac{\partial f_k}{\partial x} + \frac{\partial f_j}{\partial y} \frac{\partial f_k}{\partial y} \right) dxdy - \int_e g f_j f_k \, ds \qquad (A2.3)$$

$$b_{je} = \int_e h f_j \, ds \qquad (A2.4)$$

Some properties of these are:

(i) $a_{kje} = 0$ unless both nodes j and k are corners of the same triangle (otherwise either $f_j = 0$ or $f_k = 0$ in the entire element);
(ii) $a_{kje} = a_{jke}$
(iii) for those integrals $a_{kje} \neq 0$, the integrand of the surface integral does not depend on x, y (this is only true for linear interpolation functions).

This means that each triangular element produces only 6 nonzero contributions. Using the definitions in Fig. A2.1, the nonzero contributions are (omitting the line

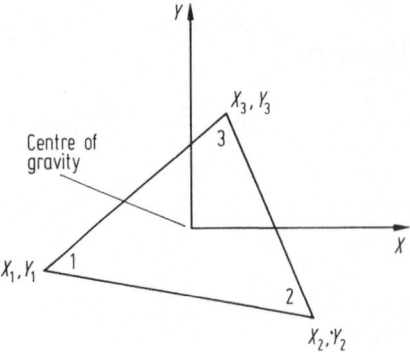

Fig. A2.1 Definitions for triangular element.

integrals for a while):

$$a_{11e} = (b_1^2 + c_1^2)/4A$$

$$a_{12e} = (b_2 b_1 + c_2 c_1)/4A$$

$$a_{13e} = (b_3 b_1 + c_3 c_1)/4A$$

$$a_{22e} = (b_2^2 + c_2^2)/4A$$

$$a_{23e} = (b_3 b_2 + c_3 c_2)/4A$$

$$a_{33e} = (b_3^2 + c_3^2)/4A \qquad\qquad (A2.5)$$

where

$$b_1 = y_2 - y_3 \qquad c_1 = x_3 - x_2$$

$$b_2 = y_3 - y_1 \qquad c_2 = x_1 - x_3$$

$$b_3 = y_1 - y_2 \qquad c_3 = x_2 - x_1$$

and

$$A = \tfrac{1}{2}\{(x_2 y_3 - x_3 y_2) + (x_3 y_1 - x_1 y_3) + (x_1 y_2 - x_2 y_1)\}$$

is the surface area of the triangle.

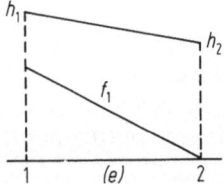
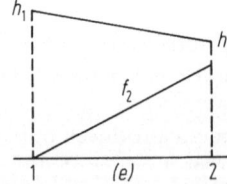

Fig. A2.2 Boundary integral (A2.4).

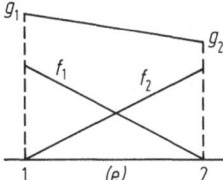

Fig. A2.3 Boundary integral (A2.3).

The line integrals in eqs. (A2.3) and (A2.4) can be computed by assuming that the functions g and h vary linearly within an element (Fig. A2.2). In each, there are contributions only from the boundary segments containing node j; in case of eq. (A2.3) moreover only if k is the same or a neighbouring boundary node (Fig. A2.3). The contributions to eq. (A2.4) are found to be

$$b_{1e} = \tfrac{1}{6}\Delta s(2h_1 + h_2)$$

$$b_{2e} = \tfrac{1}{6}\Delta s(h_1 + 2h_2) \tag{A2.6}$$

The contributions to eq. (A2.3) are (note that these should be added to those from eq. (A2.5)):

$$a_{11e} = \tfrac{1}{12}\Delta s(g_2 + 3g_1)$$

$$a_{12e} = a_{21e} = \tfrac{1}{12}\Delta s(g_1 + g_2)$$

$$a_{22e} = \tfrac{1}{12}\Delta s(g_1 + 3g_2) \tag{A2.7}$$

References

Baker, A.J. (1983) *Finite Element Computational Fluid Mechanics.* Hemisphere, New York.

Cebeci, T.&A.M.O. Smith (1974) *Analysis of Turbulent Boundary Layers.* Academic Press.

Chow, C.Y. (1979) *An Introduction to Computational Fluid Mechanics.* Wiley, New York.

Chung, T.J. (1978) *Finite Element Analysis in Fluid Dynamics.* McGraw–Hill, New York.

Courant, R.&D. Hilbert (1962) *Methods of Mathematical Physics,* Interscience, New York.

Cunge, J.A., F.M. Holley & A. Verwey (1980) *Practical Aspects of Computational River Hydraulics.* Pitman, London.

Daubert, A. and O. Graffe (1969) *Technique du modèle mathematique de l'embouchure d'un estuaire.* IAHR Congress Kyoto, pp. 479–489.

Douglas, J. and J. Gunn (1964) A general formulation of alternating direction methods, *Numer. Mathem.* **6**, 428.

Engquist, B. and A. Majda (1977) Absorbing boundary conditions for the numerical simulation of waves, *Math. Comp.* **31**(139), 629–651.

Fischer, H.B., E.J. List and R.C.Y. Kon (1979) *Mixing in Inland and Coastal Waters.* Academic Press.

Garabedian, P.R. (1967) *Partial Differential Equations.* John Wiley, New York.

Gear, C.W. (1971) *Numerical Initial Value Problems in Ordinary Differential Equations.* Prentice Hall.

Gresho, P.M. and R.L. Lee (1979) Don't suppress the wiggles, they're telling us something, in T.J.R. Hughes (ed.), *Finite Element Methods in Convection Dominated Flows.* ASME Winter Annual Meeting.

Grijsen, J.G. and C.B. Vreugdenhil (1976) *Numerical Representation of Flood Waves in Rivers.* Symp. Unsteady Flow in Open Channels, Newcastle on Tyne.

Hindmarsh, A.C., P.M. Gresho and D.F. Griffiths (1984) The stability of explicit Euler integration for certain finite-difference approximations of the multidimensional advection-diffusion equation, *Int. J. Numer. Methods in Fluids.* **9**, 853–897.

Jansen, P.Ph. (ed.) (1979) *Principles of River Engineering.* Pitman, London.

Kinnmark, I.P.E. (1986) *The Shallow-water Wave Equations: Formulation, Analysis and Application.* Springer, Berlin.

Lamb, H. (1963) *Hydrodynamics.* Cambridge University Press.

Lambe, T.W. and R.V. Whitman (1969) *Soil Mechanics.* John Wiley.

Leendertse, J.J. (1967) Aspects of a computational model for long-period water wave propagation, *Rand Corp. Mem. RM–5294–PR* Santa Monica, California.

Leonard, B.P. (1980) The QUICK algorithm: a uniformly third order finite-difference method for highly convective flows, in: K. Morgan *et al.* (eds.), *Computer Methods in Fluids.* Pentech Press, London.

Mitchell, A.R. (1977) *Computational Methods in Partial Differential Equations.* Wiley, New York.

Peyret, R. and T.D. Taylor (1983) *Computational Methods for Fluid Flow*. Springer, New York.

Pinder, G.F. and W.G. Gray (1977) *Finite Element Simulation in Surface and Subsurface Hydrology*. Nelson, London.

Roache, P.J. (1976) *Computational fluid dynamics*. Hermosa Publishing, Albuquerque, New Mexico.

Sielecki, A. (1968) An energy-conserving difference scheme for the storm surge equations, *Monthly Weather Review* **96**, 3.

Taylor, C. and T.G. Hughes (1981) *Finite Element Programming of the Navier Stokes Equations*. Pineridge Press.

Thatcher, M. and D.R.F. Harleman (1972) A mathematical model for the prediction of unsteady salinity intrusion in estuaries, *MIT Ralph M. Parsons Lab. Report* **144**, Cambridge, Massachusetts.

Thompson, J.F., Z.U.A. Warsi and C. Wayne Mastin (1985) *Numerical Grid Generation, Foundation and Applications*, North-Holland Publishing, Amsterdam.

Verboom, G.K. and A. Slob (1984) Weakly reflective boundary conditions for two-dimensional shallow-water flow problems, *Adv. Water Resources* **7**, 192–197.

Verruijt, A. (1970) *Theory of Groundwater Flow*. Macmillan, London.

Vreugdenhil, C.B. (1969) On the effect of artificial-viscosity methods in calculating shocks, *J. Eng. Math.* **3**, 4, 285–288.

Vreugdenhil, C.B. and J. Voogt (1975) Hydrodynamic transport phenomena in estuaries and coastal waters: scope of mathematical models, *ASCE Symp. Modelling 1975*. San Francisco.

Vreugdenhil, C.B. (1980) Application of finite-difference methods to estuary problems, in *Mathematical Modelling of Estuarine Physics*. Springer.

Vreugdenhil, C.B. and J.H.A. Wijbenga (1982) Computation of flow patterns in rivers, *J. Hydr. Div. ASCE*, **180** (HY1), 1296–1310.

Vreugdenhil, C.B. and N. Booij (1985) Numerical outflow boundary conditions for the shallow-water equations, *Int. J. Numerical Methods in Fluids*. **5** (5), 393–397.

Subject Index

Stochastic Hydrology and Hydraulics

ISSN 0931-1955   Title No. 477

Editors: A. Szöllösi-Nagy, Budapest, Hungary; T. E. Unny, Waterloo, Ont., Canada with an International Advisory Board

Stochastic Hydrology and Hydraulics will publish research papers, reviews and technical notes on stochastic and probabilistic approaches to hydrology and hydraulics by covering all processes of the hydrological cycle including water quality.

The contributions will encompass a wide range of theory and applications including stochastic differential equations in hydrology and hydraulics, parameter estimation and identification techniques, random hydrodynamic fields, multivariate analysis, real-time hydrologic forecasting, extreme value statistics, reservoir theory, geostatistics, stochastic control and programming, contaminant transport in random environment, wave problems, chaotic systems, stochastic turbulence modeling, stochastic boundary layer problems, and risk and reliability analysis.

Stochastic Hydrology and Hydraulics aims

● to promote the publication of manuscripts specifically devoted to the theory and application of stochastic processes in hydrology and in hydraulics by covering all processes of the hydrological cycle

● to publish in a single journal contributions by mathematicians, systems scientists, engineering hydrologists and hydraulicians, thereby promoting the exchange of ideas and the cross-fertilization of different sciences as depicted below

HYDROLOGISTS ←——————→ HYDRAULICIANS	
MATHEMATICIANS	SYSTEMS SCIENTISTS

● to further research in stochastic and probabilistic approaches to hydrology and hydraulics, thereby providing incentives to the use of such approaches in the above specified fields of engineering.

Springer-Verlag
Berlin Heidelberg
New York London Paris
Tokyo Hong Kong

Springer

Computer Methods in Water Resources

1st International Conference, Morocco, 1988

Springer-Verlag
Berlin Heidelberg
New York London Paris
Tokyo Hong Kong